SDI and Industrial Technology Policy

SDI and Industrial Technology Policy: Threat or Opportunity?

Walter Zegveld and Christien Enzing

Frances Pinter (Publishers), London

First published in Great Britain in 1987 by
Frances Pinter (Publishers) Ltd.
25 Floral Street
London WC2E 9DS

British Library Cataloguing in Publication Data
Zegveld, Walter
SDI and industrial technology policy:
threat or opportunity?
1. Technological innovations—United
States 2. Research, Industrial—United
States 3. Strategic Defense Initiative
I. Title II. Enzing, Christien
338′.06 T173.8

ISBN 0-86187-694-6

Typeset by Spire Print Services Ltd., Salisbury, Wilts
Printed by Biddles of Guildford Ltd.

Contents

List of Tables

List of Figures

List of Abbreviations

ABM	Anti Ballistic Missile
BMD	Ballistic Missile Defense
CBO	Congressional Budget Office (U.S.)
CEP	Council on Economic Priorities (U.S.)
DARPA	Defense Advanced Research Projects Agency (U.S.)
DEW	Directed Energy Weapons
DoD	Department of Defense (U.S.)
FAS	Federation of American Scientists (U.S.)
GAO	General Accounting Office (U.S.)
IC	Integrated Circuits
ICBM	Intercontinental Ballistic Missile
ISTO	Innovative Science and Technology Office (U.S.)
KEW	Kinetic Energy Weapons
METAL	Military Emergent Technology Awareness List
MITI	Ministry of International Trade and Industry (Japan)
MoD	Ministry of Defence (U.K.)
NASA	National Aeronautics and Space Administration (U.S.)
NTBF	New Technologies Based Firms
OTA	Office of Technology Assessment (U.S.)
R &D	Research and Development
RDT &E	Research Development, Testing and Evaluation
SDIO	Strategic Defence Initiative Organization (U.S.)
SET	Scientists, Engineers and Technicians
SLBM	Sea Launched Ballistic Missile
TRACES	Tracing Key Events in Science
VHSIC	Very High Speed Integrated Circuit
VLSIC	Very Large Scale Integrated Circuits

Acknowledgements

The authors would like to thank the Netherlands Ministry of Education and Science for their financial support in the preparation of a report that led to the publication of this book.

Thanks are also due to Bep Jungschlager, Ingrid Leeflang, Jeanine van der Voort and Lucienne Wibier for their patience and help in preparing the manuscript.

We would like to thank in particular Marc Zegveld for his significant effort to gather substantial and useful information during his stay in the USA in 1986.

Walter Zegveld and Christien Enzing
September 1986

1 Introduction

In March 1983, President Reagan called for an intensive and comprehensive effort to define a long term research and technology development programme with the ultimate goal of eliminating the threat posed by Soviet nuclear ballistic missiles. A number of study teams were established including the Defensive Technology Study Team, the Future Strategic Strategy Study Team and the Miller Panel. The Defense Technology Study Team, commonly referred to as the Fletcher panel, called for the structuring of a broad-based research and technology development effort focused on establishing the technical feasibility of a comprehensive ballistic missile defense system based on new emerging technologies. The recommended effort was structured to permit a decision by a future President and Congress in the early 1990s on whether to proceed with the development and the deployment of a ballistic missile defense for the twenty first century.

In January 1984, the Strategic Defense Initiative was established as a research programme based on the Fletcher Study. In the same time frame, the Strategic Defense Initiative Organization (SDIO) was formed as a Defense Agency to manage the Department of Defense effort. Specifically, a comprehensive SDI programme was defined to explore key technologies which might enable the development of defensive systems capable of intercepting ballistic missiles.

The cost of the 1986–90 SDI research programme is estimated at approximately $30 billion. The cost of an operational SDI system is estimated at between $100 and $800 billion.

Although SDI is presented as a military system, to be developed within the context of national defense, it is of strategic importance with respect to foreign policy as well and it may contain important elements of industrial technology policy. The latter position is generating much attention since the SDI research programme is being executed in a period when the strategic role of science and technology is being strongly emphasized in the industrialized world in general and in the USA in particular, with respect to both national defense and economic development.

Science and technology have become a new issue in international relations. Decrease in the more traditional forms of military and economic dominance of the major industrial countries are making these countries exercise their hegemony in the area of science and technology in their pursuit of a dominant international position. Because of the extent of SDI research and development programme, a principal question is its impact on the industrial base. In this publication we have evaluated SDI solely from a standpoint of the development of industrial technology. We have not touched upon the probably even more important aspects of the impact of SDI on national defense and foreign relations policy.

Our central thesis is that the SDI programme contains strong elements of industrial technology policy. We have approached this thesis from the perspective of the role of government in the development of science and technology and from insight in the processes of technological innovation. We have also investigated the issue of the economic consequences of stimulating scientific and technological development via the military route. For years it has been assumed that technological development in the military sector was far ahead of that in the civil sector. The supposedly large amount of spin-offs from defense and aerospace developments to the civil sector contributed to the justification of the large amounts of military and aerospace R&D-spending. However, for civil developments the military route has always been criticized and the issue of opportunity–cost has been amply discussed. Wouldn't it be much more cost-effective to choose and finance directly civil R&D goals? Dual-use technologies may change the direction of the debate as well as the potential of military spin-off from civil technology.

Our study is structured in three parts. Part I discusses the SDI programme, its origin, content, size and prospects and is covered in the chapters 2 and 3. In chapter 2 a general outline is given. This chapter is partly of a technical nature, essential to understand what the SDI-programme is and how its technical feasibility and objectives are changing in the course of the R&D and the policy development process. The SDI programme is compared with

NASA's Apollo programme, which was executed under rather similar circumstances.

Chapter 3 focusses on foreign participation in SDI. Although from the perspective of NATO-allies it would seem rather evident that these countries participate in such a voluminous defense programme, on the other hand it raises questions as to SDI's role in industrial technology development and the right of participating countries to develop and market commercial products based on SDI technology. In this chapter we give an overview of the barriers to foreign participation in SDI and the results of the negotiations between the United States and their allies on such participation. The central theme is: what is the objective of the United States inviting other countries to participate in SDI and what will be the benefit to participating countries in the development of industrial technology?

Part II, containing the chapters 4 and 5, focusses on the role of government in technological development and on the (meso) process of technological innovation. Both chapters start with a more general theoretical introduction and lead to the question whether SDI be (partly) considered as military or industrial technology policy.

Chapter 4 introduces the general theme of industrial technology policy. It sets out the development of technology policy and the role of government in this process. A balanced and integrated industrial technology policy is stated to contain three interrelated elements determining innovative potential: technological opportunity, structure of the industrial sector and the size and structure of market demand. Defense-supported R&D is, however, with a growing budget, playing an important role in the direction and the pace of technological innovation in United States industry.

In chapter 5 we focus on the policy character of the SDI programme. Is it a defense R&D programme, or is it a substitute for an industrial technology programme? To answer this question, it is important to look at the composition of the SDI research budget. If there is substantial civil, commercial, spin-off from military projects, then civil and military technological trajectories will more or less converge. The type of research not directed to applications, neither civil nor military, and for which results might be of dual use, is basic research. The question then is: what is the ratio basic/applied research and what are the conditions for universities to perform defense sponsored basic research? Will the diffusion of the knowledge generated within SDI take place to other sectors of the economy without constraints? Another element with respect to the defense or industrial policy character of R &D sponsored in the military context is to look at the technologies which are being developed or considered to be relevant for the military sector. What can be the conclusion when we compare the core technologies that the Pentagon is stimulating with the programs of the Japanese Ministry of International Trade and Industry?

The second part of our study on the more macro-economical level of industrial technology policy is followed by the third and last part which is focussing on the process of technological innovation and the economic significance of military R &D-spending (chapters 6 and 7).

In chapter 6 a description is given of the innovation process and the successive models that try to catch the essence of the conditions that define the success of technological innovation. What is the contribution of basic scientific research to technological development and to successful innovation? What is the nature of scientific and technological progress? Does scientific understanding precede technological advance, or vice versa? Do technological concerns shape the scientific enterprise? The classical question in economics studies of 'push or pull' is touched upon. All these more theoretical questions on the role of scientific and technological development in the innovation process, and on their influence on economic development, can directly be applied to the discussion of the character of the SDI-programme.

Chapter 7 focusses on the economic benefits of SDI. The ultimate argument in the debate on the economic legitimation of military R &D — and in this special case of SDI-funding — is the spin-off argument. It is the key to understanding the defense-economy debate. In this chapter we sketch the spin-off debate in particular with regard to SDI. What is the relation between the opinions of the several participants in the debate and their position and interest in the debate? What are the arguments used in the spin-off debate and how can spin-off be measured? Is there a relationship between government R &D-spending and economic growth?

Chapter 8 summarizes and draws conclusions with respect to the role of SDI in the development of industrial technology.

Part I: The Programme

2 SDI: objectives and extent of the programme

In his address to the nation on March 23, 1983, President Reagan advocated that the United States 'embark on a program to counter the awesome Soviet missile threat with measures that are defensive'. He called upon the American scientific community, 'those who gave us nuclear weapons, to turn their great talents now to the cause of mankind and world peace, to give us the means of rendering these nuclear weapons impotent and obsolete'. The President chose a surprising way to present his vision. In the closing moments of a televised speech arguing for a larger military budget, Reagan abruptly changed topics and announced what was no less than a revolution in American strategic doctrine. Mr. Reagan had drafted the closing paragraphs of the speech himself only five days before it was televised and they were seen in advance by only a handful of advisors.

With these statements President Reagan appealed to the scientific community to devise new methods for countering the threat of nuclear ballistic missiles and revived the public debate concerning strategic defense against nuclear weapons which seemingly was laid to rest by the 1972 ABM treaty. The resurgence of interest which resurfaced in the early 1980s, was brought about in part by technological advances in a number of different fields: new possibilities for weapons systems employing laser and particle beams and kinetic energy projectiles; along with advances in guidance and sensing systems, optics, and computer processing. These advances seemed to make ballistic missile defense a more attractive and realistic proposition than it had been in the 1960s and 1970s.

It was not until ten weeks after this speech that Mr. Reagan appointed three panels of experts to assess the feasibility of achieving the goal he had already set out. These panels were appointed to assess the technological prospects and security implications of strategic defenses, their mandate was focussed only on ballistic missile defense (BMD).

Three panels

The Defensive Technology Study Team (DTST — known as the Fletcher panel, after its chairman, James C. Fletcher) reviewed over a period of four-and-a-half months the above mentioned emerging technologies, relevant to ballistic missile defense. They also examined concepts for fashioning these technologies into a robust, reliable system. The Fletcher panel opened the body of its summary report with a rhetorical question: 'What has happened to justify another evaluation of ballistic missile defense as a basis for a major change in strategy?' The answer in the report the panel said that although enormous hurdles remain, the technological advances of the past two decades show great promise for BMD. The new technologies offer the prospect of reaching out and intercepting ballistic missiles almost from the moments they are launched. To give some shape to the President's general goal of comprehensive missile defense as a focus for research, the Fletcher panel sketched out a hypothetical, multi-tiered defense system based on these prospective technologies, with each tier engaging missile warheads at a different phase of flight.

The panel recommended that a vigorous research and development program should be pursued. The goal of that program would be to demonstrate the key technologies by the early 1990s, allowing a decision to be made at that time on whether to proceed with a ballistic missile defense for the twenty first century. The other two panels examined the policy impact of new defense technologies for United States strategic objectives, and the effect this defense might have on the relationship with the allies and the Soviets.

The report of the Miller panel (after its chairman Franklin C. Miller, director of Strategic Forces Policy in the Pentagon) has not been publicly released. The other policy group, the Future Security Strategy Study Team (the Hoffman panel, after its chairman, Fred S. Hoffman), did issue a declassified summary of its study. The Hoffman panel proposed progressive stages of defense deployments, starting most immediately with anti-tactical ballistic missile (ATBM) defenses for Europe, based upon available technologies for terminal defense. More or less anticipating that the United States would never reach the ultimate goal of nearly perfect defense, the Hoffman panel's report confined itself to being principally a brief for stabilizing retaliatory deterrence. The panel's proposals for partial defenses are considerably less ambitious than those of the Fletcher group, still less ambitious than the SDI. At the same time the panel served to broaden the rationale for the SDI by widening the issues beyond simply 'Star Wars' technologies.

The conflicting advice of the Hoffman and Fletcher panels was not confronted, debated and then integrated in a strategic defense policy. Long after the panels finished their work, the issues of a less-than-perfect rather than nearly perfect defenses, defenses of retaliatory forces rather than cities, a world free of mutual nuclear threats rather than one of buttressed deterrence were still not placed in strong focus. This failure is considered to rest with the nature of the panels themselves. From the beginning, they were contending with a president who had already committed himself publicly to a goal of nearly perfect defenses. Although members of the panels clearly believed that more modest expectations were warranted, they had no reason on that point alone to confront Reagan and force him from his commitment. The panel members were in fact quite content with more modest goals than 'perfect defense', and thought that even modest goals justified a vigorous BMD research program. Moreover, either by design or good fortune, the defense system and research program sketched out by the Fletcher panel also seemed to discourage debate and hard decisions rather than to provoke them. Almost everyone's favorite BMD technology or strategic purpose could find a place within the multitiered defense system (Hafner, 1985).

The ballistic missile trajectory: four phases

The Fletcher recommendations formed the basis for a new Strategic Defense Initiative Organization, which was almost immediately established in the Defense Department. This new organization did pursue a technology development program which was very similar to the recommended one.

The various systems concepts of the new program are designed to counter a massive, full-scale Soviet attack, which, at current force levels, would involve thousands of intercontinental and submarine-launched ballistic missiles and the tens of thousands of warheads that they carry.

To understand the various defensive concepts, it is necessary to know something more about the typical trajectory of a ballistic missile. The flight of a land-based intercontinental ballistic missile (ICBM) would follow a trajectory over the Arctic from one superpower to the other, last 25 to 30 minutes, and cover approximately 10,000 kilometers. When equipped with multiple warheads (MIRVs), the ICBM would have four phases of flight: a boost phase, during which large rocket boosters accelerate the missile to a velocity of seven kilometers per second; a post-boost phase, during which the missile deploys its warheads and decoys; a mid-course, where the warheads and decoys follow a ballistic trajectory through the vacuum of space; and a terminal phase, during which the warheads re-enter the atmosphere (the decoys having been burned up, or slowed by friction) and descend toward their targets (see figure 2.1).

The *boost phase* spans the first several hundred seconds after launch, when the missile is lifted from its silo and thrusts through and out of the atmosphere by its first, second, and third-stage booster rockets, propelling the missile at increasing speeds to an altitude of about 200 kilometers, the altitude of the lowest earth-orbiting satellites. By the end of the boost phase, the missile is travelling at seven kilometers per second.

Missiles in boost phase are particularly vulnerable to destruction by a BMD system for two reasons. First, the missile booster is a large and fragile target, easily damaged by directed energy weapons (lasers, particle beams) and kinetic energy projectiles. Second, the flame emitted by the booster rockets give off intense infra-red radiation, providing space-based defense satellites with a fairly accurate fix on the missile's location. This same infra-red radiation is what provides certain United States and Soviet satellites with early warning of an enemy attack. There is a clear and powerful advantage to destroying ICBMs in boost phase, as destroying a single missile will result in the destruction of all the warheads and decoys on that missile. Given that current ICBMs and SLBMs carry anywhere from three to fourteen warheads and might carry one hundred decoys, successful boost-phase intercept can give a BMD system important leverage over the offense by substantially reducing the number of warheads and decoys that will confront the mid-course and terminal defense systems.

Once the third-stage booster burns out, the missile enters the

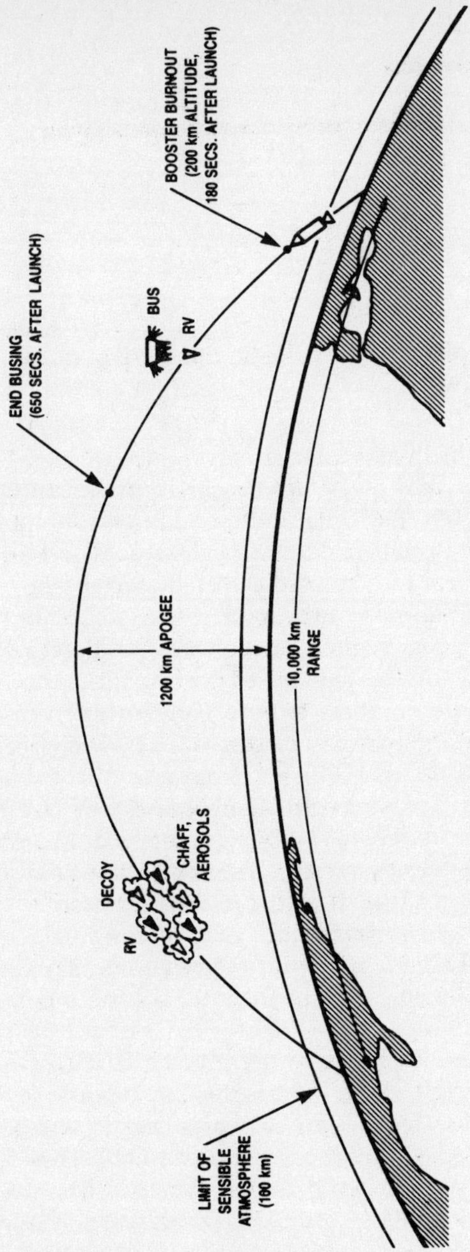

Figure 2.1 Phases of flight for an ICBM with booster characteristics similar to the U.S. MX missile
Source: Bethe, Boutwell and Garwin, 1985.

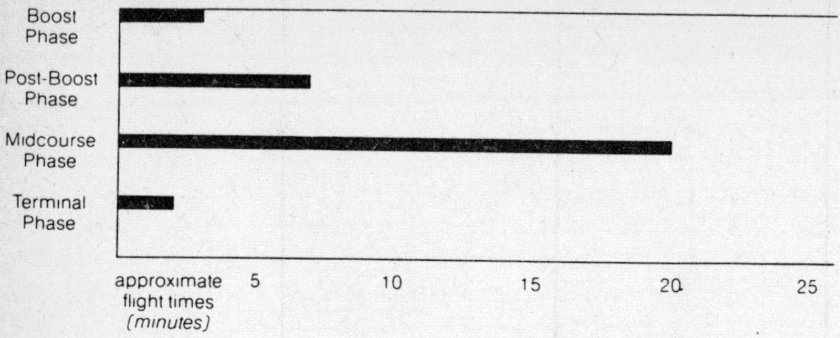

Figure 2.2 Four phases of a ballistic missile trajectory
Source: Fletcher, 1984.

post-boost phase, which lasts another several hundred seconds. The third-stage booster falls away, leaving the post-boost vehicle and its cargo, up to ten multiple independently targeted re-entry vehicles, or MIRVs, each carrying a nuclear warhead directed at a separate target. Powered by a low-thrust rocket, the post-boost vehicle, or 'bus', maneuvers through space, dropping off its re-entry vehicles in programmed sequence and directing them on their distinct trajectories. The post-boost vehicle can also carry decoys and other penetration aids to confuse or overwhelm the defense; hundreds of these may be deployed by each booster along with each re-entry vehicle. Once released from the bus, the re-entry vehicles and decoys begin their ballistic, or freefall, flight, arching up to their apogee at 1,000 kilometers or so, then falling back to earth. This twenty minute unguided ascent and descent of the re-entry vehicle is known as the *midcourse phase*.

In addition to its nuclear warhead, each re-entry vehicle contains sufficient heat shielding to allow it to survive re-entry through the atmosphere and a fusing system to detonate the warhead at the right time.

The use of decoys to increase greatly the number of possible targets is especially effective, because in the vacuum of space the trajectories and velocities of warheads and decoys will be indistinguishable. These decoys could consist of suitably shaped pieces of metal or plastic, or aluminized balloons, which will appear to BMD sensing systems as actual warheads. In addition, the release of fragments of metal wire, called chaff, or clouds of aerosol could hide the actual warheads by overwhelming the BMD sensing systems.

A more fundamental problem with mid-course interception concerns how to locate precisely these warheads and decoys in the first place. Unlike missile boosters with their infra-red signatures, warheads and decoys do not give themselves away to space-based satellite sensors. A BMD system must either search for the warheads and decoys, which emit modest amounts of infra-red radiation and reflect the earth's faint infra-red light, or it must illuminate these objects by reflecting radar waves or lasers off them.

The *terminal phase* begins when the re-entry vehicles, decoys, penetration aids, and debris begin to re-enter earth's upper atmosphere at an altitude of about 100 kilometers. Lightweight objects, including some decoys and debris are slowed by atmospheric drag and then broken apart by the force of deceleration. The heaver re-entry vehicles, which are 'hardened' to survive heat and deceleration, continue on their trajectories. The terminal phase ends some two minutes after it begins when the re-entry vehicles, by this time a glowing red, detonate over their targets. Advances made since the 1960s in radars and other sensors, computer processing, and the accuracy of BMD interceptors has made terminal defense a much more viable proposition than it was twenty years ago. In particular, striking improvements in radar and computer technology now make it possible to use non-nuclear interceptors instead of the nuclear warheads (see Fletcher, 1984; Berthe, Boutwell and Garwin, 1985).

An effective defense against a massive Soviet ballistic missile attack will require a multilayered system capable of attacking missiles in all phases of their trajectory. Each of these multiple layers, corresponding to one of the four phases of a missile's flight, would use its own combination of sensors and weapons to trade and intercept missiles as they streak through space. Each layer serves as a back-up to the previous layer: those missiles that penetrate or 'leak' through the first line of defense, the boost-phase defense, will be intercepted by past-boost, mid-course, and then terminal-phase defenses. By contrast, the earlier anti ballistic missile (ABM) concepts attacked oncoming missiles primarily in their re-entry phase (figure 2.3).

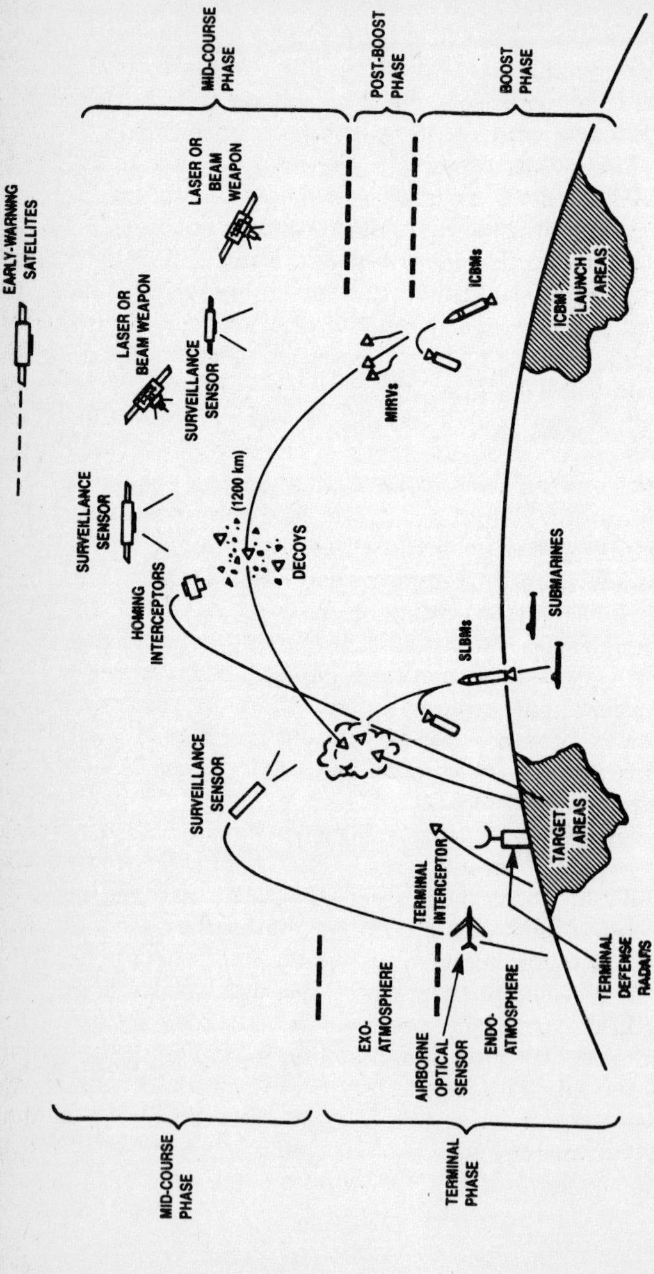

Figure 2.3 Multilayered BMD system
Source: Bethe, Boutwell and Garwin, 1985

New BMD weapons technologies

The purpose of the Strategic Defense Initiatives is to conduct research on those technologies which might enable the development of defensive systems capable of intercepting ballistic missiles after they have been launched and preventing them from hitting their targets. Much of the public focus on SDI has been centered on the promise held out by new BMD weapons technologies, especially the so-called directed energy weapons. The other four technical areas which are funded by the Department of Defense as SDI projects are kinetic energy weapons; surveillance, acquisition and tracking; systems analysis and battle management; and support programmes.

Directed energy weapons (DEW)

These include various types of laser and particle beams, which travel at or near the speed of light (300,000 km. per second). Creating a promising DEW candidate will be a challenge; it will involve developing a laser and/or particle beam of sufficient intensity and accuracy to deposit large amounts of energy on small areas of rapidly moving ballistic missiles and warheads from great distance.

The term 'laser' is an acronym, standing for *l*ight *a*mplification by the *s*timulated *e*mission of *r*adiation. Essentially, lasers can provide very narrow beams of coherent light with sharply defined wave-lengths. Four types of lasers are currently under consideration for BMD missions: infra-red chemical lasers, electrically driven ultraviolet lasers, free-electron lasers, and X-ray lasers produced by nuclear explosions. These laser beams could destroy a missile booster or a warhead in one of two ways. Chemical, ultraviolet, and free-electron lasers (all known as optical lasers, being similar to visible light rays) would use a concentrated beam of light to burn into the missile 'skin'. The X-ray laser works differently, penetrating only a very thin layer of the booster skin, but initiating a shock wave that could destroy the missile's internal components or knock it off course. Particle beams are produced by particle accelerators (atom-smashers). If the particles (hydrogen ions, for example) are accelerated to nearly the speed of light, they can penetrate deeply into material targets. Yet charged particle beams are not suitable for transmission over large distances because they are deflected by the earth's magnetic field. Scientists have instead proposed using neutral hydrogen beams, which might be obtained by accelerating negative hydrogen ions and stripping off the extra electron. Particle beams have an advantage over lasers in that their deep penetration into the target could destroy the electronics and guidance systems of the MIRV bus. Another advantage is that the beam from the accelerator can be steered by electric and magnetic fields, without moving the accelerator itself; this is important if several targets are to be attacked in quick succession.

The DEW program brings together the directed energy efforts addressing four basic concepts identified by the Defensive Technologies Study Team as potentially workable: space-based chemical lasers, ground-based eximer and free electron lasers, space-based particle beams, and space-based X-ray lasers. Research will focus on demonstrating integrated systems that include beam generators; beam control; large and adaptive optics; and acquisition, tracking, and pointing. Its emphasis will be on new and innovative approaches.

Kinetic energy weapons (KEW)

These are typically rocket-propelled weapons that accelerate small projectiles. These projectiles 'home in' on the target using infra-red sensors for guidance and tiny rocket thrusters to change direction.

These projectiles can destroy a booster or warhead simply by high velocity impact. The velocity of the missile and/or warhead actually contributes to its destruction; the projectile hurtles towards its target either head-on or at right angles, penetrating it to a depth many times its own length. The weapons themselves would be carried by a large satellite, and fired by either rockets or, possibly, electromagnetic railguns.

Goals for the KEW program include designing non-nuclear warheads and fusing, high performance interceptor missiles, and fire and navigational controls necessary for kinetic weapon ballistic missile defense (BMD) systems. This project will explore 'novel and advanced' kinetic weapons concepts such as hypervelocity guns as well as developing existing BMD technologies. Kinetic weapons are being researched for use in space as well as terminal defense.

Systems analysis and battle management

This program is divided into two projects. One aims at developing technologies necessary for a 'highly responsive, ultra reliable, survivable, endurable and cost effective' battle management system necessary to direct a low-leakage defense system. The other will provide overall guidance to weapons, sensors, and supporting technologies through extensive systems analysis. A major part of the battle management effort will be to quantify the technological risks and costs of achieving such a complex system. The systems analysis will include threat assessment, mission analysis, and system evaluation. The most immediate efforts will aim at achieving high performance processors and software, and responsive, secure communications networks that provide high reliability and fault tolerance. Evaluation and demonstration of this complex system will largely depend on early development of modelling and simulation tools (see Marlin, Lippin, 1985).

Surveillance, acquisition and tracking

This program will examine technologies to support the decision to initiate a defensive engagement, manage the battle, and assess the status of forces during a battle against nuclear ballistic missiles. Its major goal is to develop signal and data processing technologies for discriminating threatening re-entry vehicles from other objects and backgrounds. Radar and infra-red technologies will be explored.

Support programs

The fifth program will focus on two major questions: how can the United States scientifically predict the minimum energy required to destroy Soviet systems in all possible engagement scenarios? What technologies and concepts will be necessary to protect a deployed ballistic missile defense system from a dedicated attack by the offense? Additional support programs will fund development of multimegawatt power systems and space logistics capabilities.

These five technical areas, based on the Fletcher recommendations formed the basis of the Fiscal Year 1985 and 1986 budget requests for the SDI. Figure 2.4 gives an overview of the SDI-budget for the five programs and of the growth of the whole SDI-budget as a percentage of the total national R &D.

The strategy changes

It was recognized that the limited time and resources available to the Fletcher Panel provided only an initial look at the best approach to the problem. Thus the SDI Organization has initiated a three-phase series of Architecture Studies to refine the research priorities of the program. Each successive phase of these studies involves a progressively smaller number of teams working at a higher level of effort. The result is expected to be a greatly improved understanding of the fundamental issues and choices posed by the SDI, and a much better sense of the proper focus for the research and development effort.

Phase I of the Architecture Studies involved ten teams working for a period of about six months. These studies were completed in the late summer of 1985, and subsequently five of the teams were selected to proceed into the Phase II studies which will continue through 1986. Ultimately a single contractor will be selected for

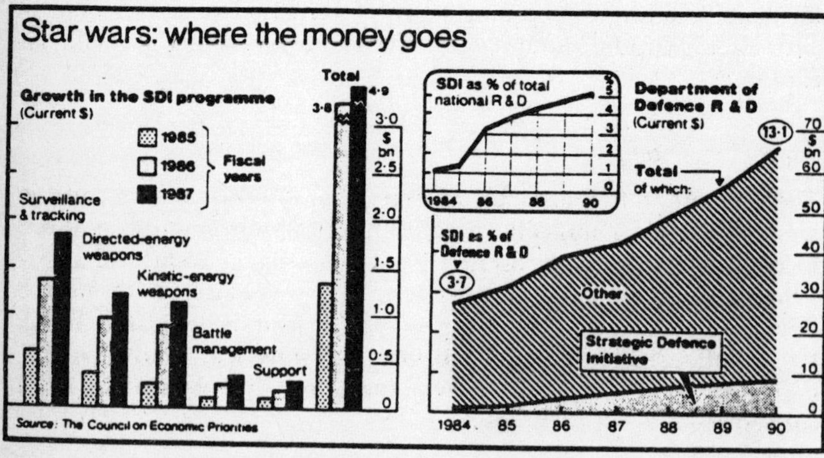

Figure 2.4 Star Wars budget, as percentage of total national R &D
Source: The Council on Economic Priorities

Phase III, which will continue for the duration of the SDI program. The Phase III contract will include a clause that will prohibit the contractor from working on other aspects of the SDI program. This hardware exclusion clause would place the System Architect in the position of an impartial referee for the rest of the program.

The Phase I Architecture Studies have produced a vision of the SDI that is radically different from that envisioned when the SDI program was initiated. The various changes to individual projects are the result of basic conceptual changes. Instead of four or five layers each with a twenty per cent leakage rate, up to seven layers are called for, with an average leakage as bad as forty per cent. The original program plan was looking at a single architecture that had a number of space-based components. Now the SDI is looking at two architectures: one with a few space-based elements, and one with no space-based elements. The original architecture omitted any special plans for Europe, but this oversight has been corrected.

However, the goal of the program is the same: a highly effective but not perfect defense that in the event of war would limit damage to the United States in conjunction with other offensive weapons such as the Trident II and the New Design Attack Submarine (Pike, 1986).

In a staff study conducted at the request of senator Proxmire, senator Chiles and senator Johnston the authors attempt to provide some perspective on the progress achieved at the time the Strategic Defense Initiative nears its third anniversary. The findings of this study are that while some significant progress has been achieved in each of the five major programs of the Strategic Defense Initiative, none of it could be described as 'amazing'. Interviews with key SDI scientists involved in the research revealed that there have been *no major breakthroughs* which make a mid- to late-1990s deployment of comprehensive missile defenses more feasible than it was three years ago.

In fact, the 'schedule-driven' nature of the current research program, which requires that a development decision be made by the early 1990s, has aroused significant concern among scientists at the national weapons laboratories. The concern is twofold: first, promising long-term research will be compromised to reach an arbitrary schedule; and secondly, in an effort to maintain public support for high funding levels and an early development decision, SDI experiments will degenerate, in the words of a senior scientist at the Livermore National Laboratory, into 'a series of sleazy stunts'.

Much of the progress that has been achieved has resulted in a greater understanding of program difficulties, which are much more severe than previously considered. The authors are struck by myriad uncertainties and unknowns at every turn in the program — uncertainties and unknowns that bear directly on the effectiveness of a comprehensive ballistic missile defense that might be deployed in future. Much of that uncertainty is likely to remain, for even with strategic defenses in place, the United States will never be able to test the system adequately under realistic conditions (Waller, Bruce and Cook, 1986).

According to SDI chief scientist Gerold Yonas the SDI-strategy has changed: 'lasers based in space are now a "backup" option'. Research will emphasize lasers on the ground, where their enormous energy supplies can be more readily situated. Using and replacing energy for tests is also more economical for ground-based lasers. The excimer and free-election lasers have been the favoured ground-based models. Neutral-particle beam, X-ray and chemical lasers were planned for space. The SDI concept has, except for congressional cuts in the SDI's budget also been refined by predictions of Soviet counter measures.

The Soviets might reduce the launch phase when the rockets are firing, from 300 to 100 seconds, so a faster weapon would be needed — lasers instead of orbiting canons that propel small objects at 20 or 30 kilometers per second with electromagnetic pulses. These orbiting railguns present the same problems of a bulky power supply and vulnerability as space-based lasers. Orbiting infra-red sensors were thought to be the best way to track missiles and to discriminate between real bombs and decoys after they are ejected from their boosters. Designed for the 20-minute mid-course phase, this space surveillance and tracking system (SSTS) is in a 'process of rethinking' Yonas says (see Joyce, 1985).

The cost of SDI*

If President Reagan's vision of a space-based system for defense against ballistic missile attack ever becomes a reality, its dollar cost

*Based on the report 'The SDI: costs, contractors and consequences', Council on Economic Priorities, 1985.

would be high. Some estimates by official sources reach as high as $500 billion. Other Administration spokespeople dismiss such estimates as being without foundation. Since planners have not even begun to define the architecture of a strategic defense system, they say, no valid estimates of cost are possible.

Although the full price cannot be predicted accurately, a preliminary estimate can be obtained by projecting production costs from initial research and development funding. A completed strategic defense system might cost roughly $400 to $800 billion. These figures are based on the general historical relationship between pre-full scale development (FSD) expenditures and total production cost. This educated guess assumes the present, five-year program will immediately lead to FSD. With the uncertainty surrounding the technologies, however, pre-FSD spending could continue for a number of years driving total program costs higher. All that can be said with certainty is that the Defense Department plans to spend some $32.2 billion over the next six years on research and development of strategic defense systems (Abrahamson, 1984).

In an economy producing more than $3 trillion worth of goods and services each year, a $32 billion six-year program is not likely to bankrupt the United States. Nonetheless, this kind of investment involves a significant opportunity cost. Money could be spent in many other areas. The attempt to shield the SDI program from public scrutiny was cast aside when the Pentagon dropped the crucial criterion that a proposed Star Wars defense must be 'cost-effective at the margin' if it is to be built. Economic implications of SDI depend both on the level of funding and the marginal spending increase involved in the project. The total spending level indicates the amount of resources, both human and capital, involved in the project. The marginal increase measures the new demand for resources the program will make on the economy.

Sizing the SDI-budget

Funding the strategic defense had grown prior to the President's speech. Budget authority for SDI-related program elements grew by 81 per cent in constant dollars between 1980 and 1983 (Pike, 1984). In 1985 and 1986, though, SDI funding is projected to 'take

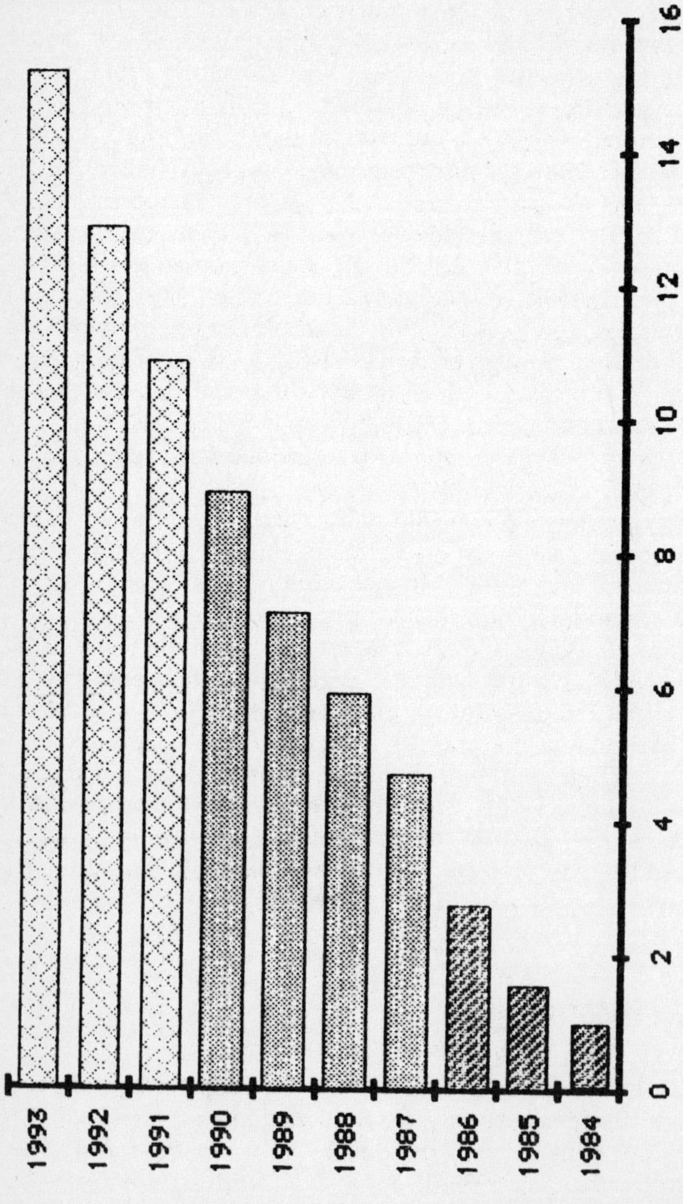
1993
1992
1991
1990
1989
1988
1987
1986
1985
1984
0
2
4
6
8
10
12
14
16

The Research Costs of SDI

Year	Cost ($mil)
1984	992,000
1985	1,538,000
1986	2,759,000
1987	4,742,000
1988	5,926,000
1989	7,118,000
1990	8,926,000
1991	10,892,000
1992	12,899,000
1993	15,206,000
	70,998,000

Notes:

1984–86: Funds appropriated by Congress

1987–90: Present funding requests by the SDI organization

1991–93: Projected funding requests

Figure 2.5 Research costs of SDI in billions of 1986 dollars

Source: The Federation of American Scientists, SDI-budget & Program Study, 1986

Note: Includes only the Department of Defense portion of the SDI budget off'. Appropriations have increased by 36 per cent in 1985 alone. During 1986, budget authority will more than double in real terms.

Since some growth was already planned, new funding resulting from the President's initiative does not become a large part of the budget until 1986 (see Table 2.1). Indeed, as a result of Congressional budget actions, SDI funding for 1985 is $130 million lower than previously planned. In 1986, new funds will account for 51 per cent of the total.

Lack of unclassified, long-range budget plans makes it impossible to precisely determine the overall increase in strategic defense funding resulting from SDI. Two estimates, however, have been offered. One, presented by SDI project manager Lt. General James Abrahamson, suggests DoD plans formulated prior to the Initiative called for spending roughly $15 billion on strategic defense research between 1985 and 1989. Thus, the SDI would add some $9 billion to research on defensive systems. After adjusting for inflation, the increase would be between $7 and $8 billion (in 1985

Table 2.1 New budget authority for SDI (millions of dollars)

	1984	1985	1986	1984–6 Total
1985 DOLLARS				
Funds from Existing program elements	$941	$1,527	$1,809	$4,277
New Funds	51	−130	1,912	1,833
Total	$992	$1,397	$3,721	$6,110
CONSTANT 1986 DOLLARS				
Funds from existing program elements	$1,015	$1,588	$1,809	$4,413
New Funds	55	−135	1,912	1,831
Total	$1,070	$1,453	$3,721	$6,245

Source: Congressional Budget Office, 'Analysis of the Costs of the Administration's Strategic Defense Initiative, 1985–1989' (Washington, D.C.: CBO, May 1984), p. 5. Updated using data from R, D, T, & E Programs (R-1), Department of Defense Budget for Fiscal Year 1986, February 4, 1985.

dollars). The other estimate, an independent assessment of strategic defense funding, performed by John Pike of the Federation of American Scientists (FAS), found that additional funding attributable to SDI amounts to over $11 billion between 1985 and 1989 (see Table 2.2). This is about $10.3 billion after adjusting for inflation.

A judgement of the validity of either estimate, however, is suggested by comparing these estimates to the detailed CBO figures for the year in which the three sources overlap: 1986. In that year, the SDI Office claims $2,600 million of the proposed $3,721 million budget was already planned. FAS estimates the amount of pre-planned strategic defense funding at $2,233 million. The CBO, on the other hand, could find only $1,809 million programmed prior to the President's speech. This suggests that neither the SDI Office nor the FAS estimate captures the total increase due to the new effort, though the FAS estimate is closer to the mark.

After three years, the SDI budget has nearly tripled. SDIO has slowed the pace of some of its research efforts, although this has not been done primarily as a result of Congressional budget cuts, as SDIO officials have claimed. Decisions to downgrade certain research efforts seem to be driven as much by their lack of technical promise (as was the case with chemical lasers), and by SDIO's insistence on keeping to an unrealistic 'technology-limited' research schedule. The FY (fiscal year) 1987 budget request for the SDI clearly includes a considerable margin of cut insurance. This was also true of the FY 1986 request of $3.7 billion, which was $1.2 billion over the $2.5 billion that the SDIO actually wanted. In the submission of the FY 1986 budget the SDIO indicated that the FY 1987 request could be about $4.9 billion. By November 1985 the planned FY 1987 request had been reduced to $4.7 billion, as a result of delays in the program and down-selection of various technologies. In December 1985 the planned request was further reduced to $4.5 billion.

At the last minute, over $0.5 billion was added to the request to provide extra insurance against Congressional cuts in the program. This was confirmed by a 'senior White House official' who was quoted in the *Los Angeles Times*. The large number of projects that have been reduced in scope and pace suggests however, that the SDIO's actual goal for FY 1987 is only $3.3 billion (Pike, 1986).

There are three remarkable trends visible in the SDI budget. We

Table 2.2 Increase in SDI budget authority 1984–9 (Millions of dollars)

	1984 $	1985 $	1986 $	1987 $	1988 $	1989 $	1985–9 Total
1985 DOLLARS							
SDI Funds	992	1,397	3,721	4,910	6,166	7,300	23,494
Pre-SDI Funds							
Directed Energy	317	409	444	488	531	596	2,469
BMD	636	1,168	1,789	1,925	2,305	2,608	9,795
Increase	39	−181	1,488	2,496	3,330	4,096	11,229
CURRENT 1986 DOLLARS							
SDI Funds	1,070	1,453	3,721	4,701	5,649	6,418	21,942
Pre-SDI Funds							
Directed Energy	342	426	444	467	487	524	2,348
BMD	686	1,215	1,789	1,843	2,112	2,293	9,252
Increase	42	−188	1,488	2,390	3,050	3,601	10,342

Source: John Pike, Strategic Defense Budget (Washington, D.C.: Federation of American Scientists, 1984). Updated using data from RDT&E Programs (R-1), Department of Defense Budget for Fiscal Year 1986, Febr. 4, 1985.

already mentioned the shift in priorities within the budget that the SDIO makes every year and the increase in the annual budget. The third trend is the fact that the SDI's budget will not be cut as a result of the Gramm-Rudman-Hollings Act. This Act forces the federal government of the United States to cut its deficit. This Gramm-Rudman-Hollings Act is declared partly nonconstitutional, including SDI. Because of this highly controversial act, the government will have to cut all budgets in all departments, except the SDI-budget 'despite the very tight fiscal environment'. President Reagan said that when he sent his budget plans for 1987 to Congress, 'the budget provides for maintaining, and in some cases expanding, high priority programs in crucial areas of national interest'. Those crucial areas include, among others 'the space-program, non-military research and national security'.

SDI and Apollo

The debate on the economic benefits of R &D spending and as a focus of our study of military R &D, especially SDI, can learn something by comparing it with an earlier experience of similar magnitude.

While some have mentioned the Manhattan project as a parallel, the effort to build the atomic bomb took place during wartime, when the economy was tightly controlled. Thus from an economic standpoint, it would not tell much about the likely consequences of the SDI. The National Aeronautics and Space Administration (NASA) and the Apollo program is a better example. Coming during a period of strong economic growth and high military spending, the Apollo effort occurred under somewhat similar circumstances to those facing the SDI.

During the peak year, 1966, NASA as a whole and its Apollo program in particular took up 20.8 and 13.5 per cent of all research and development spending in the United States, respectively (see Table 2.3). SDI will not be nearly so large, at least until it gets closer to the production phase. SDI RDT &E could consume 5.1 per cent of total R &D during 1990. Relatively speaking, SDI will use less than half the R &D resources required for Apollo.

In marginal terms, however, SDI is of similar magnitude. During the two years of peak growth in the Apollo effort, 1965 and 1966, the project required 40 and 20.8 per cent of the new R &D resources

Table 2.3 NASA and Apollo research and development outlays (Millions of dollars)

	1961	1962	1963	1964	1965	1966	1967	1968	1969
Total NASA[1]	$744	1,549	2,568	4,038	4,354	4,538	4,282	4,035	3,272
Apollo Program[1]	—	—	—	2,273	2,615	2,941	2,923	2,556	2,025
Total R &D[2]	$14,316	15,394	17,059	18,854	20,044	21,846	23,146	24,605	25,631
NASA %	5.20	10.06	15.05	21.42	21.72	20.77	18.50	16.40	12.76
Apollo %	—	—	—	12.06	13.04	13.46	12.63	10.39	7.90
Federal R &D[2]	$9,250	9,911	11,204	12,537	13,012	13,968	14,395	14,928	14,895
NASA %	8.05	15.63	22.92	32.21	33.46	32.49	29.74	27.03	21.96
Apollo %	—	—	—	18.13	20.09	21.05	20.30	17.12	13.60

Sources: 1. Federal Budget, Fiscal Years 1963–71 (Washington, D.C.: U.S. GPO, 1962–70).
2. Science Indicators 1982 (Washington, D.C.: National Science Board, 1983), Appendix table 2-3, p. 235.

(see Table 2.4). SDI could use 34.6 per cent of all new R &D funds during 1986. Another similarity is that both programs began after sustained growth in space (for Apollo) or defense (for the SDI) funding that consumed a substantial share of new R &D-spending (see Figure 2.6).

Prior to the Apollo program, NASA did not significantly compete with private industry for scientists and engineers (S &E's) even though its workforce tripled from 3,500 to 8,200 S &E's between 1960 and 1962. Still, as the Apollo program geared-up in 1963, members of Congress began to express concern about the space program competing with private industry for technical manpower (GPO, 1963). NASA almost doubled its total S &E employment during that year. In the peak year of the Apollo program, 1966, NASA employed almost 92,000 scientists and engineers. This workforce was 5.5 per cent of the national S &E's total. SDI could employ 12,000 S &E's during 1987 (roughly 5 per cent of the national total). Several studies suggest that in a period of rapid increase in space-related demand for scientists and engineers, other industries had to match the traditionally-high aerospace and electronics industry salaries to slow occupational transfers. Benefits of space research and development, however, extended beyond salary considerations, and NASA contractors were in a

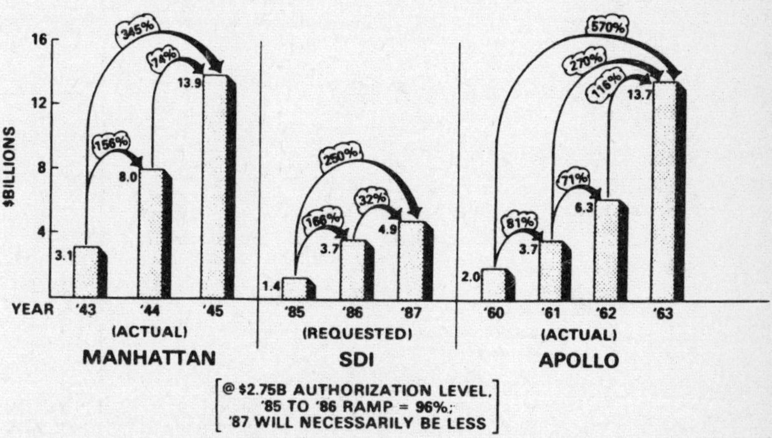

Figure 2.6 Comparative funding (in 1985 dollars) SDI-growth less than comparable national priority programs

Source: Hearing on SDI, House of Representatives 1985

Table 2.4 Increase in NASA and Apollo funding (Millions of constant 1985 dollars)

	1961	1962	1963	1964	1965	1966	1967	1968	1969
Total NASA	$2,691	5,514	8,811	13,514	14,570	14,734	13,426	12,164	9,377
Last Year	$1,459	2,691	5,514	8,811	13,514	14,570	14,734	13,426	12,164
Increase	$1,232	2,823	3,297	4,703	1,056	164	−1,307	−1,263	−2,787
Apollo Program	$ —	—	—	7,800	8,750	9,548	9,165	7,706	5,804
Last Year	$				7,800	8,750	9,548	9,165	7,706
Increase	$ —	—	—	7,800	950	798	−384	−1,459	−1,902
Total R&D	$51,757	54,802	59,605	64,701	67,082	70,929	72,581	74,178	73,462
Last Year	$49,210	51,757	54,802	59,605	64,701	67,082	70,929	72,581	74,178
Increase	$ 2,547	3,045	4,803	5,096	2,380	3,847	1,652	1,598	−716
NASA %	48.36	92.70	68.65	92.28	44.38	4.26	−79.13	−79.03	389.13
Apollo %				153.05	39.93	20.75	−23.23	−91.31	265.55
Federal R&D	$33,442	35,283	39,147	43,023	43,548	45,351	45,140	45,005	42,691
Last Year	$31,798	33,442	35,283	39,147	43,023	43,548	45,351	45,140	45,005
Increase	$ 1,644	1,841	3,864	3,876	524	1,803	−211	−135	−2,313
NASA %	74.91	153.33	85.31	121.34	201.50	9.09	619.26	935.24	120.47
Apollo %	—	—	—	201.24	181.31	44.26	181.78	1,080.5	82.22

Source: Figures from previous table converted to 1985 dollars using Defense Department deflators for RDT&E outlays. National Defense Budget Estimates for FY 1985 (Washington, D.C.: OASD (Comptroller), March 1984), p. 62.

much better position than most civilian industries to adapt to the tight market during the mid-1960s. Indeed, the present dilemma facing the United States basic industries may find part of its roots in the competition with space and defense work for highly-skilled labor than began in the 1960s.

As an example of the impact of the Department of Defense on the scientific and technical community, it is interesting to note the employment statistics of scientists and engineers resulting from its activities. Using its Defense Economic Impact Modelling System, the Department generated the following figures for Congressional testimony in March 1983. (see Table 2.5). These figures do not include SDI activities.

While in macroeconomic terms the Apollo effort at its height was clearly a much larger undertaking than the research phase of SDI, the new demands SDI will make on the economy are already of similar magnitude. Given the nature of the strains accompanying the Apollo effort, there is reason to think SDI could impose painful costs on industries requiring substantial numbers of engineers and other technical personnel.

If the SDI proceeds into full scale development and procurement, it would most likely swamp the Apollo program in an economic sense. As Dr. Richard DeLauer, Under Secretary of Defense for Research and Engineering, testified back in November of 1983:

> I participated in the Apollo program, and a lot of blood, sweat and tears and a lot of my hair went into making that engine work. It was going to land a guy on the Moon, and we made it work. I'm not saying that the Apollo program isn't significant. I'm saying that this [the SDI] is very very significant. This is a multiple of Apollo programs because of the environment in which the systems will have to be deployed.

Contractors

Bidding for strategic defense contracts is already intense. Over 240 companies submitted bids for contracts to define the 'system architecture' for strategic defense that were awarded to 10 firms in December 1984. These are the top ten contractors ranked by

Table 2.5 Employment in thousands as a result of defense and other activities

	1981			1987		
	Defense	Other	Total	Defense	Other	Total
Aero-astronautic engineers	24.03	41.07	65.10	40.62	46.54	87.16
Chemical engineers	2.73	49.35	52.08	3.79	52.16	55.96
Civil engineers	12.59	143.46	156.04	15.48	166.39	181.87
Electrical engineers	45.14	261.72	306.86	61.19	298.56	359.75
Mechanical engineers	27.21	168.62	195.83	39.17	188.14	227.31
Metallurgical engineers	2.50	14.8	16.68	3.48	16.63	20.10
Engineers (other)	35.85	351.68	387.53	53.28	406.42	459.70
Atmospheric and space scientists	0.70	11.15	11.85	1.04	11.90	12.94

Biological scientists	1.24	60.88	62.12	1.52	68.88	70.40
Chemists	6.21	116.90	123.11	8.84	130.10	138.94
Geologists	0.85	31.19	32.04	1.30	37.18	38.48
Marine scientists	0.48	4.60	5.08	0.60	5.14	5.75
Physicists and astronomers	6.55	17.71	24.26	7.57	18.75	26.32
Life and physical scientists (other)	1.34	6.18	7.52	1.48	5.50	6.98
Mathematicians	3.06	6.94	10.00	3.41	7.30	10.71
Statisticians and actuaries	1.39	31.91	33.30	2.24	37.84	40.08
Computer specialists	57.15	399.76	456.91	71.29	475.75	547.04
TOTALS	229.02	1717.30	1946.31	316.30	1973.18	2289.49

Source: Department of Defense, United States

amount of contracts award (in thousands of dollars):

Lockheed: homing overlay anti-satellite weapon, Talon Gold laser pointing system, space-based ICBM interceptors and surveillance systems.	$531.966
TRW: chemical, free-electron and excimer lasers, computers and surveillance.	$323.862
Boeing: airborne optical systems for tracking, lasers, rail guns and projectiles.	$217.394
Teledyne Brown: engineering and technical support, computers and software.	$179.993
Rockwell International: semiconductors, excimer laser, optical sensing, rail guns and their power supplies.	$165.714
Hughes: sensors, optics, beam directors, rail guns and chemically powered interceptors, lasers, computers, system architecture.	$155.655
RCA: radar, computer components.	$125.702
McDonnell-Douglas: terminal defense, computer memories, interceptors, laser booster, system architecture.	$122.301
LTV Aerospace: interceptor rocket, rail gun power supply.	$ 98.638
AVCO: excimer laser, X-ray laser.	$ 72.660

These top 10 contractors took-in almost $500 million in 1985, about one-third of the budget. Large aerospace companies stand to make the most, although the SDI office has set aside about one per cent of its budget for small business. Many of the projects now funded under the SDI office began years ago. Laser weapons, rail guns and chemically powered rockets are a few examples. Some companies had already started projects on their own initiative. What the SDI office has done is to organize these projects. The top ten-companies have gained an inside track in the battle to determine both the type of strategic defense system that may ultimately be deployed and in receiving future contracts for the components of such a system. The active role of the contractors in defining the need for and capabilities of a strategic defense system that could yield them billions of dollars over the next two decades raises serious questions of potential conflict of interest. CEP points out the danger that exists that SDI may develop an economic momentum that will make it difficult to stop the program even if it becomes clear, on strategic and technical grounds and on cost-effectiveness, that the United States would be better defended without deployment of a strategic defense system (Marlin, Lippin, 1985).

References

Abrahamson, Lt. General, J., *Statement on the President's Strategic Defense Initiative before the U.S. Congress*, Senate Appropriations Committee, May 1984.

Bethe, H.A., Boutwell, J., Garwin, R.L., 'BMD Technologies and Concepts in the 1980's', *Daedalus Vol. I: Weapons in space: Concepts and Technologies*, Spring 1985.

Fletcher, J.C. 'The Technologies for ballistic missile defense', *Issues in Science and Technology*, The National Academy of Science, Washington D.C., Fall 1984.

Hafner, D.L., 'Assessing the Presidents' Vision', *Daedalus Vol. 1: Weapons in Space: Concepts and Technologies*, Spring 1985.

House of Representatives, *Impact of Strategic Defense Initiative on the U.S. Industrial Base*, Hearing before the subcommittee on Economic Stabilization, Committee on Banking, Finance and Urban Affairs, December 1985.

Joyce, C., 'Industry takes a shine to star wars', *New Scientist*, p. 18, December 12, 1985.

Marlin, A.T., Lippin, P., *The Strategic Defense Initiative: Costs, Contractors & Consequences*, Council on Economic Priorities, 1985.

Pike, J., *The emperor's newest clothing: changes to the SDI as a result of Phase I — architecture studies*, Federation of American Scientists, Washington D.C., February 16, 1986.

Pike, J., *Strategic Defense Budget*, Federation of American Scientists, Washington DC., 1984.

Union of concerned Scientists, *Issue Backgrounder: Star Wars: Myth versus Reality*, Myth 6: The SDI is affordable, June 6, 1986.

U.S. Congress, *Nation's manpower Revolution*, Senate Committee on Labor and Public Welfare, Washington, D.C. U.S. GPO Part 8, pp. 2690–9, November, December 1963.

Wallern, D., Bruce, J., Cook, D., *SDI: progress and Challenges*, Staff Report Submitted to Senator Proxmire, Senator J. Bennett Johnston and Senator Lawton Chiles, March 1986.

3 SDI: Foreign participation

On March 26, 1985, United States Defense Secretary Caspar Weinberger, addressing NATO's Nuclear Planning Group in Luxembourg, explicitly invited the NATO nations, as well as Australia, Israel and Japan, to participate in the SDI research program. At the conclusion of their meeting, the NATO defense ministers issued a unanimous declaration of support for SDI research, but NATO leaders continue to express their concern that research should be conducted within the terms of the ABM Treaty. This invitation fits into NATO strategy, long familiar in defense debates on both sides of the Atlantic, that better value for the defense dollar and a more effective defense system would both result from a greater harmonization of the military hardware produced by the members of the NATO alliance. We might interpret the above invitation in terms of international NATO security, if serious concern had not been expressed about the off-setting benefits that may be provided to foreign governments and corporations in what some perceived as an attempt to win political support for SDI. There is a growing fear in Washington that a technologically strong, but independently oriented Europe would not necessarily suit the best interests of the United States.

In a hearing on the 'Impact of the Strategic Defense Initiative on the United States industrial base' for a Subcommittee of the House of Representatives, December 1985, Chairman LcFalce remarked that

> it appears that the administration is willing to pay other governments and foreign corporations to conduct research for SDI and allow other countries and foreign corporations the right to control and market commercial products which are based on the technology developed by this U.S.-financed research. The impact of such policies on the U.S. high-tech research capability, the competitive position of U.S. high technology industries, and U.S. employment and production could be substantial and warrants careful examination.

This last remark underlines the technology policy character of the SDI-programme and it is within this perspective that we discuss foreign participation in SDI.

Forms of co-operation

An amendment to the 1986 Defense Department authorization bill, proposed in 1985 by Senator Sam Nunn, earmarks $200 million to spend only on collaborative R &D projects between United States companies and those from other members of NATO. The reaction on this proposal in Europe has been mixed. Some countries have been quick to sign up for the six projects that had already been put forward, but at the same time fears have been raised that a United States-led program could undermine Europe's own attempt to start rationalizing its defense industry through joint research efforts.

The result of Weinberger's invitation has been a series of continuing bilateral discussions with many of the countries; briefings to their delegations who have come to Washington; arranging visits for these groups to United States laboratories and contractor facilities; and, in some cases, visits from SDIO technical teams abroad. The object of these dialogues has been to address the various procedural concerns and to identify areas of SDI research for possible participation, consistent with United States law, with security interests in protecting sensitive technology, and American obligations under the ABM Treaty.

In these discussions, the United States indicated flexibility with regard to mechanisms enabling participation. There are a number of ways that the allies could participate in the program. One way would be in concluding new Government–to–Government overarching, or framework agreements which would provide a comprehensive basis for the participating country's involvement. Proceeding along these lines would allow the United States to sort out the issues of principal concern regarding foreign participation. Matters such as ensuring fair property rights, providing proper access to SDI information, maintaining strict industrial and technical security, and allowing fair competition with United States industry are among these concerns. Whether or not the United States will conclude such overarching agreements, specific research projects will require separate implementing arrangements: these could take a variety of forms. One form could be an arrangement whereby the United States would enter into an agreement for co-operative research, in which both parties would share the costs. The implementing vehicle in this case would be a Government–to–Government Memorandum of Agreement. Such co-operative arrangements can be particularly attractive and have potential since there are many cases where the United States and an ally are both working in comparable areas.

A more common form of participation would be a contract awarded from the SDIO (or its agent) to an allied Government, or Government Agency, sometimes on a sole source basis, but more normally on a competitive basis. The allied government or government agency might, in turn, sub-contract to its industry or laboratories. Allied governments, in this instance, would play a management, or integration role. SDI could also award contracts directly to allied industry, again on a sole source or competitive basis, as appropriate. Finally it is believed there is considerable latitude for allied laboratories and industry to work directly with the United States' concerns on SDI contracts.

It has been pointed out that the fundamental basis for allied participation must be technical merit and there can be no guarantee of achieving a certain level of effort.

One final possibility has been indicated. Allies, particularly those with more limited national research capabilities, could be invited to send selected and appropriately cleared personnel who can make real contributions, to join specified United States research teams in a scientific exchange program. The Department of Defense (DoD) already has a number of such exchange programs in place. Whatever the arrangement, the United States is taking vigorous steps to assure the security of the program and to safeguard all sensitive information and technologies. This is fundamental to any agreement that may be reached.

Barriers to foreign participation

Regarding foreign participation in SDI and especially with respect to the barriers limiting this participation, much can be learned from a statement by Mr. John Pike, Associate Director for Space Policy of the Federation of American Scientists. This statement was made in the Hearing on the 'Impact of SDI on the U.S. Industrial Base', Subcommittee On Economic Stabilization. Mr. Pike in his statement mentioned that there are at least six current barriers to significant foreign participation in SDI. Mr. Pike's testimony is quoted below:

Barrier 1 — The anti-ballistic missile (ABM) treaty of 1972

The first barrier to the participation of European corporations in the SDI is the Anti-Ballistic Missile Treaty of 1972. Almost half of the SDI program consists of work on the development and testing of anti-missile components limited by the ABM Treaty. In addition to limits on the testing and deployment of these components, the Treaty also specifies that: '. . . each party undertakes not to transfer to other states . . . ABM systems or their components limited by this Treaty. (Article IX.)'

The Treaty further defines this to include the obligation: '. . . not to provide to other States technical descriptions or blue prints specially worked out for the construction of ABM systems and their components limited by the Treaty.' (Agreed Statement G.)

The Treaty limitations apply only to ABM components that currently consist of interceptors and radars, and to components based on other physical principles capable of substituting for these components in the future. However, the Treaty does not place limitations on ABM computers, and thus would pose no barrier to cooperation in this area. But the Treaty limitations on the transfer of ABM components and plans for such components would effectively preclude European participation in such major SDI projects as space-based, air-based, and ground-based sensors, space-based and ground-based rocket interceptors, and directed energy weapons. This prohibition would even apply to the Low Endo-atmospheric Defense Interceptor (LEDI), which has been proposed for defence against tactical missiles in Europe, since LEDI is also being developed to defend American ICBM silos against strategic missiles, and is thus subject to the ABM Treaty's limitations. It is difficult to imagine how non-U.S. corporations could join in these projects without access to 'technical descriptions or blue prints'.

These companies could perhaps work as subcontractors, building very small devices and subsystems, but the total dollar value of such work would be trivial. Although European companies could perhaps perform basic research in areas such as directed energy weapons, they would be precluded from following up their initial research efforts with more substantial (and lucrative) development and testing contracts. Although the United States might benefit by skimming the cream of European industry, European companies and governments would probably find this an unattractive prospect.

Although the Administration has shown a casual disregard for the letter of the ABM Treaty, adherence to the so-called 'restrictive' interpretation of the Treaty would also include compliance with the Article IX and Agreed Statement G restrictions. There has been no indication that yet another reinterpretation of the Treaty is in the offing.

All European nations, but especially France and Great Britain, have a strong interest in preserving the ABM Treaty and have staunchly supported the Treaty since its inception. The ABM Treaty prevents Soviet deployment of an ABM system that, while ineffective against United States strategic forces, could be more useful in protecting the U.S.S.R. from British and French retaliation. European leaders, notably British Foreign Minister Sir Geoffrey Howe, have made compliance with the ABM Treaty a key condition for their support of the SDI. Even if the American government is prepared to sacrifice the Treaty on the altar of European participation in the SDI, it is far from clear that European governments will go along with the process.

Barrier 2 — The limited capabilities of European companies

The demanding requirements of defense against ballistic missiles will place additional limitations on European participation in the SDI. As much as one-third of the SDI budget, over $10 billion, will go for research into technologies with which European companies have little expertise. The United States has been doing work on ABM systems for almost 30 years spending over $50 billion during this period. This has created a substantial industrial base in anti-missile technology in the United States that is lacking in Europe.

The European competitive disadvantage is illustrated by the almost total absence of non-United States firms in ABM and SDI contracting to date (see figure 3.1). Since President Reagan delivered the 'Star Wars' speech on 23 March 1985, over 1,000 contracts worth more than $3 billion have been signed. All of these contracts are with American companies. The only case of non-United States participation in an SDI contract to date is a small subcontract that Plessey in the United Kingdom has signed with Boeing. A review of Army ABM contracts over the decade 1975–85 reveals a similar number of contracts with about the same value. Not a single Army prime contract was awarded to a European company.

Since the ABM Treaty was signed in 1972, there have been literally thousands of contracts worth perhaps $10 billion signed with hundreds of companies in the United States. We have been able to identify only two very small contracts, worth no more than $10 million, that were awarded to European companies, both of which, ironically, are French. The Navy bought a 2 meter diameter mirror from RESOC (Recherches et Etudes d'Optique et de Sciences Conexes) and the Army shipped a sample radar antenna face to France to be tested in a solar furnace for vulnerability to nuclear explosions.

Although tactical European military systems are the equal of many of their American counterparts, expertise in tactical weapons does not necessarily carry over to strategic weapons. European experience with small short-range tactical radars provides little basis for working with very large phased array radars capable of tracking thousands of missile warheads at distances of thousands of kilometers. European work on small heat-seeking missile guidance components with dozens of sensor elements has little application to the production of anti-missile sensors with millions of sensor elements. The guidance and propulsion technology developed for short-range rockets by European companies falls far short of the requirements posed by anti-missile interceptors, whose acceleration, speed and range must all be over an order of magnitude greater.

Several countries, such as the Federal Republic of Germany, do have small programs for the development of directed energy weapons. However, these efforts lag behind the United States by over a decade. Recently plans were unveiled by one German firm for a mobile anti-aircraft laser mounted on a tank chasis. It was stated that several years would be required to complete development of the initial test model. The United States first tested such a system in the mid-1970s.

Source	Period	All contracts total value $	European Contracts $	European Share %
SDI	1983–5	3,000,000,000	0	0.0
Army BMD	1975–85	3,000,000,000	0	0.0
SDI & related	1972–85	10,000,000,000	10,000,000	0.1

Figure 3.1 European participation—the record
Source: Hearing on the Impact of Strategic Defence Initiative on the U.S. Industrial Base for the Subcommittee on Economic Stabilization, December 1985.

The United States imposes strict limitations on the export of advanced military systems through such means as the 'Military Critical Technology List', which includes virtually the entire SDI. These technologies are subject to strict licensing and controls to prevent their export to Soviet block countries. This process places significant limitations on the ability of Allied countries to obtain access to these technologies, and would impede European collaboration in the SDI. Special clearances are needed for European nationals to gain access to United States facilities, and the June 1983 Report of the Defense Science Board Task Force on Industry-to-Industry International Armaments Cooperation noted that:

> Cumbersome and lengthy security procedures are a significant impediment to technical exchange and cooperation . . . Bureaucratic and overly protective security procedures can thwart the best intentions to cooperate by imposing time delays and by rendering the technical exchange required for industrial cooperation difficult, or even impossible.

The limited capabilities of European companies also have implications for contracting procedures, which are governed by Department of Defense Directive 2010.6 (5 March 1980) 'Standardization and Interoperability of Weapons Systems and Equipment within the North Atlantic Treaty Organization.' This directive notes that non-United States contractors would not be allowed to pre-solicitation and other briefings preparatory to awarding contracts:

> In any case when it can be established that the expertise residing in firms in (other) countries is not sufficiently advanced to allow a reasonable expectation of winning a competition . . . informal discussions should be initiated with representatives of these countries in the U.S. Often, such discussions will result in agreements that such acquisitions are not suitable prime contract opportunities for such foreign contractors.

This would seem to apply to much of the SDI.

There are additional procedural difficulties. Some are relatively minor, such as the fact that announcements of contract competitions are only sent to companies that the contracting agency believes are qualified to bid on the contract, with all other interested parties having to go to the office of the agency to get a look at the announcement. Other restrictions are more serious. European optical companies may have a hard time bidding on SDI contracts since the 1983 determination by the Defense Department that the

American optical industry was in a state of decline, and in need of protection. The application of the 'Buy American' policy to the optical industry led to the virtual exclusion of non-United States optical firms from Defense contracts.

Barrier 3 — the commercial potential of some SDI technology

The SDI is unlikely to have significant commercial benefits, but to the extent that it does, European companies are unlikely to receive contracts with spin-off potential. The SDI will be developing technologies that are irrelevant to commercial applications, under conditions of strict military secrecy. Most of the SDI research with civilian applications is already under development by civilian companies. There is in fact a limited demand for some of these exotic technologies, either from the commercial or military sectors.

Nonetheless, advocates still claim that the SDI will produce a number of commercial spin-offs. To the extent that these claims are believed, it is unlikely that the Congress would look favorably on using SDI money to subsidize European competition for American companies. Some Europeans speculate about the possibility of SDI contracts aiding the development of their computer industry. This is not a likely result of the SDI, but to the extent that it might be, the American political system would be unlikely to tolerate such a development.

Thus in the areas of computer hardware and software, as well as space transportation and other areas with the potential for commercial development, European corporations are unlikely to obtain significant SDI contracts. This is also true for the basic research efforts of the SDI's Innovative Science and Technology Office, since it is unclear that the SDI should be fostering the progress in basic science in other countries.

This does not mean that there will be significant civilian spin-offs from the SDI, or that the SDI could be supported on the basis of such spin-offs. But much has been made of the potential commercial value from the SDI, and advocates have recalled the space program and early military contributions to civilian industry. In fact, these precedents suggest that the SDI is unlikely to have significant civilian benefits. The military did lead the way in the introduction of commercial jet aircraft and miniaturized computers, but these contributions were based on factors irrelevant to the SDI. The Air Force bought large numbers of KC–135 tankers, enabling Boeing to reduce the production costs of the 707 passenger jet. The Minuteman ICBM program was the first major customer for electronic chips, enabling the electronics industry to lower costs and improve production techniques. But in both of these cases, the impact of the military was as a purchaser of production items. The SDI research program will not make this type of contribution.

The United States spent over $75 billion on the Apollo moon program alone, but the major commercial spin-offs were Tang, Velcro, and Corning Ware. The moon program was not justified on the basis of these spin-offs, nor should the SDI be justified in this way.

Indeed, the SDI may have a negative effect on the American economy. At the macro-economic level, the growth of the SDI budget will complicate efforts to reduce the Federal deficit. Economists have long noted that countries with high levels of military spending generally have lower growth rates than those with lower military burdens. In a sense the SDI is part of a trend toward re-making the American economy along the lines of that of the Soviet Union, at a time when increasing numbers of business leaders recognize the need to emulate the Japanese model of lower military spending and higher economic growth.

In some areas of interest to the SDI, such as supercomputer programmers and optical engineers, shortages of trained personnel already exist, and these would be exacerbated by the SDI. The SDI's short-term infusion of large amounts of cash may drive up the salaries of some favored technicians, but it is unlikely to expand the pool of talent. Instead, scientists and engineers will move from more beneficial endeavors to work on the SDI.

The specific economic benefits that have been claimed for the SDI are greatly exaggerated. To begin with, the SDI will stress the development of performance parameters that have limited relevance to the civilian world. Star Wars computers must be able to survive the effects of nuclear explosions, and so the SDI is putting money into research on radiation-hardened Gallium Arsenide computer chips. But banks and insurance companies don't need computers that can continue working during World War III. Directed energy systems can also have industrial and medical uses. But the military needs weapons with power outputs of many millions of Watts that will operate for a few minutes, whereas civilian tools must operate for months or years at powers of only a few Watts. X-ray lasers could be used in medical research or the production of computer chips, but the SDI is working on X-ray lasers powered by nuclear explosions, and no hospital or factory is going to detonate atomic bombs in its basement on a regular basis.

Another difficulty with finding commercial applications for SDI technology stems from the classification restrictions on military research. Much of the SDI will be highly classified, and thus unavailable for commercial development. Precedent for such problems is found in the

Defense Department's Very High Speed Integrated Circuit (VHSIC) program, which has developed new computer chips that are much faster than existing models. But the program is so highly classified that companies working on VHSIC are prohibited from even publishing photographs of the new chips.

In addition, many of the technologies that are claimed to have commercial applications are already in development by civilian companies. Computers are frequently mentioned as one area of potential spin-off to the commercial sector. But the Defense Technology Study Team (the Fletcher Panel) noted in its report on SDI computing requirements that the computer speeds required for missile defense would be available in civilian computers in the 1990s and that there was no need for the SDI to develop its own computers. There are seven commercial Gallium Arsenide computer chip production lines in the United States in addition to the one run by the SDI. Several companies not involved in the VHSIC program have produced commercially available equivalents of these high-speed chips, and have published pictures to prove it. Although the SDI has expressed an interest in the development of ultra-fast optical computers that use photons of light rather than electrons to store and process information, but there are already a number of civilian researchers at major laboratories working in this area.

The SDI spin-off argument also assumes a supply side model of technological innovation in which the introduction of a new technology creates a new market. In fact, in many cases, the introduction of new technologies is largely governed by market demand. Great reductions in the cost of space transportation is also seen by some as a potential commercial contribution of the SDI. But [at the time of the decision on SDI] there [was] already an over-supply of space launch services, indicated by the intense competition between the Shuttle and the European Ariane. American companies failed in their effort to find a commercial market for existing expendable launch vehicles, and have turned to the Defense Department as a customer of last resort. Lowering the cost of placing satellites in orbit would have a marginal impact on the commercial development of space. The main problem facing the space industry today is the lack of demand. The market for communcations satellites has not expanded at predicted rates, and there is a glut of these satellites. Recently Mexico launched its Morelos communications satellite on the Shuttle, even though the satellite will not be placed in operation until 1989, because it was cheaper to store it in space than in a hanger on the ground. Materials processing and resource monitoring in space have failed to develop as viable industries, not because of high launch costs, but because of the lack of market demand.

One additional indication of the limited commercial potential of the SDI can be seen in the limited military applications of some of these projects. Arcane technologies that are unable to find military missions are unlikely to fare better in the civilian sector.

Many of the projects in the SDI are not new weapons, but have in fact been under development for many years, although for applications other than missile defense. However, these systems were far too advanced for these other applications, and had failed to receive approval for actual development. In a sense, the SDI has become a technological orphanage. By incorporating these projects into Star Wars, with its formidable operational requirements, these systems have gained a new lease on life.

The Space Based Imaging Radar Project is the Star Wars incarnation of the Air Force Space Surveillance Technology Project. Research on the technology involved in this project has been under way for a number of years. However, very high costs, uncertain performance, platform vulnerability, coupled with the availability of more cost effective alternatives, have precluded deployment. The Navy, which considered deploying such a system under the Clipper Bow project, has since been reoriented into the less ambitious Integrated Tactical Surveillance System (ITSS).

The Imaging Laser Project is based on the previous work done on the use of Blue-Green lasers for detection of submarines by the disturbances they cause in the optical scattering layer of particulate debris that is located several hundred feet under the ocean's surface. However, after several years of work, it became clear that this technique was not very useful for detecting submarines.

The Booster Surveillance and Tracking System is the new name for the Advanced Warning System. However, during 1984 the Air Force decided that the improved resolution afforded by the technology under development in the Advanced Warning System project was far in excess of that required for the early warning mission. Very accurate determination of boost-phase trajectories was determined to provide little improvement over existing systems in warning of launch or impact point prediction.

The Space Surveillance and Tracking System was formerly known as the Space-Based Surveillance System. This system was originally intended to support the new Air Launched Miniature Vehicle ASAT, as well as more advanced high altitude ASAT systems. However, less costly near-term radar upgrades were determined to provide a more cost effective alternative.

This is not to say that European participants in SDI research might not pick up some skills that could be turned to other purposes. But the possible development of commercial or non-strategic military products cannot be used to justify the SDI for the United States. However, it is clear that the SDI is unlikely to be turned into a second Marshall Plan for the renovation of European industry.

Barrier 4 — some of the work must be performed in this country

About $1 billion of the SDI consists of activities, such as management support, that by their nature must be performed in the United States. Non-United States companies are unlikely to bid on such contracts, and would be unlikely to receive these contracts even if bids were made.

This estimate of $1 billion is conservative, since many other contracts, particularly those dealing with directed energy weapons, are for support of activities at United States facilities. For example, most of the work on Free Electron Lasers is done at Livermore and Los Alamos National Laboratories, and it might be difficult for European contractors to make a significant contribution to these efforts. These contracts would also have to be executed in the United States and European companies would be at a severe disadvantage.

Barrier 5 — the competitive environment

Given all these barriers, there is only about $1 billion of the SDI that European companies would be eligible to bid for. This estimate of $1 billion is based on the list of 18 areas of opportunity that are reported to be part of the Memorandum signed by the British government. However, these contracts will be eagerly sought by many qualified American companies as well. And in many cases the American company will enjoy various competitive advantages. Thus it would be surprising if European companies won as much as a third of these contracts, at most $300,000,000 over a five year period. In reality, the total could be much less.

Barrier 6 — the penalties of foreign preferences

Recognizing the many barriers facing European companies, the SDI has established at least two areas where non-United States contractors are virtually guaranteed a piece of the action. Several European contractors will be paid to study possible architectures for anti-tactical missile defense of NATO. The total value of these studies is unlikely to exceed $5 million, and their principal value will be to develop a more credible sales pitch for Star Wars in Europe.

The most visible SDI contract in Europe will be the award in the spring of 1986 of a contract worth perhaps $10 or $20 million for support of SDI sensor experiments on a Shuttle flight in 1987. This is part of a special set-aside in the SDI budget for international programs. A total of 33 bids have been received, including five from European companies and two that are joint proposals. Sperry has teamed with the German firm of Dornier to offer Dornier's Instrument Pointing System (IPS) which recently flew on the Shuttle as part of the Spacelab D-1 mission. Boeing has teamed with the German firm of Messerschmitt-Boelkow-Blohm (MBB), which is proposing the use of the SPAS free-flying satellite to carry the SDI sensors. Lockheed is also proposing a stripped-down version of its TALON GOLD pointing and tracking system to support this mission.

Although it is likely that one of the German/American teams will win the competition, the award may prove controversial. The Boeing/MBB proposal would provide a platform for the sensor experiments, but it would not answer the critical question of whether the sensors can be isolated from the vibration of spacecraft, since the SPAS satellite would float free of the Shuttle. The Sperry/Dornier IPS could provide answers to this question, but the IPS performed poorly in its first flight, and the Spacelab equipment it requires would take up most of the Shuttle's payload bay. Lockheed has two points in its favor. The TALON GOLD device is much more compact than the IPS, and its selection would keep together the Lockheed design team. But its lack of a European partner may weigh against Lockheed.

This suggests that contract provisions that give preference to European contractors may disadvantage American firms, as well as the rate of progress in the SDI research program.

Conclusions

The lack of publicly available details on the recently signed Memorandum of Understanding with the British makes it difficult to draw firm conclusions on the future of collaboration in the SDI. To the extent that this and future agreements have not broken the barrier to extensive participation, European support for the SDI and the Administration's current posture in the Geneva arms talks will soon decline. These barriers are not the result of bureaucratic caprice of historical accident, but of considered national policy. A summary of the consequences of barriers to foreign participation is given below (see figure 3.2).

In his statement before the Congressional Committee Mr. Pike also said the following:

While European companies will probably get $30 million in contracts fairly soon, they are unlikely to get more than $300 million over the next five years, and will almost certainly fail to realize their expectations of $3 billion. The political repercussions of these failed expectations cannot help

	$
Total SDI Program 1986–90	30 Billion
Contracts constrained by:	
1—The ABM Treaty	14 Billion
2—Capabilities of European companies	10 Billion
3—Commercial potential of SDI	4 Billion
4—Work performed in this country	1 Billion
Total not available to European companies	29 Billion
5—Available for competition	1 Billion
6—Foreign preference programs	30 Million

Figure 3.2 Barriers to European participation in SDI
Source: Hearing on the impact of strategic defence iniative on the U.S. industrial base for the subcommittee on economic stabilization, December 1985.

but further reduce European support for the SDI and the Administration's current posture at the Geneva arms control negotiations.

The alternative is a significant effort by the Administration to change the rules to facilitate European participation in the SDI. The very limited publicly available information on the Memorandum signed with the British does not allow an assessment of whether such changes are contemplated. But if special procedures are established for the SDI, the Congress should carefully examine these to determine their consistency with established policies in this area. (House of Representatives, 1985).

It was mentioned by Mr. Pike in his testimony that there has already been some informal discussion in Congress of possible measures in response to the Memorandum of Understanding with the British. These suggestions have been quoted to include:

— Special set-asides for Allied contractors could be prohibited;

— Evaluations of contract proposals could be prohibited from using bonus points for the inclusion of Allied subcontractors;

— Issuance of Requests for Proposals to companies outside the United States could be limited to only those non-United States firms that have facilities that cannot be duplicated in the United States;

— All prime contracts could be limited to American companies, unless there is a finding that a non-United States company has a unique capability that does not exist in the United States;

— The United States Government should retain property rights in the results of developments by non-United States companies that are the same as those applied to domestic firms;

— Allied countries could be required to contribute an amount equal to the value of the SDI contracts to a fund that would be used for contracts with American firms.

SDI participation

The different forms of foreign participation in SDI we mentioned at the start of this chapter are now illustrated in this section. Here we give an overview of the state of participation in the SDI program by Japan, the United Kingdom, France, the Netherlands and the German Federal Republic and some other countries.

Japan

The Japanese government has long refused to comment about the possibility of participating in SDI. The Americans issued the invitation to join in the project in early 1985, and the United States defense secretary, Caspar Weinberger, was in Asia in April 1986 to try to convince Japan of the benefits of participation. The Japanese were in two minds over the issue. On the one hand they worried that if they did not join the SDI, they would be left behind technologically. On the other hand, they did not want to get caught in a web of secrecy if they do join the project. In addition the Japanese were also concerned about becoming involved in a military project in space.

Despite such fears, prime minister Nakasone, has in September 1986 accepted the United States' invitation.

Nakasone is reported to believe that it is 'only natural' that a nation which strongly depends on the United States for its defense should express its support by joining. However, Nakasone has long been waiting for the recommendations being drafted by members of a fact-finding mission to the U.S., which returned in April. The team toured facilities in the United States being used for SDI-related research. Engineers from 21 Japanese companies participated in this mission, which was the third that the Japanese have sent. Nakasone's main forseeable problem was that of the Japanese constitution, which the United States drew up after the Second World War, and which prohibits the export of arms, and a resolution by Japan's parliament in 1969 which condemns the use of space for military purposes. Nakasone's government has, decided that parliamentarian approval is not necessary and participation in the SDI by private companies could be covered both under the Japan–United States Mutual Security Treaty and the agreement concluded in 1983 on the exchange of military technology.

The Japanese government had already cleared the way legally for participation in the SDI by making a bill to tighten up counter-espionage measures. In the face of determined opposition, this bill had to be aborted. Opposition to the bill reflected fears among Japanese scientists that the Americans may attempt to use SDI contracts as a means of controlling the flow of technological information. Japan has an open information policy which on occasions, upsets the defense establishment in the United States. For example, scientists at Osaka University's fusion laser facility, second in size only to the United States' equivalent, claim that the Americans keep begging them not to publish their results. Companies such as Sony, a participant in a SDI mission to the United States, are reported to worry that they might be prevented from producing consumer products out of technology acquired through participating in SDI.

In Tokyo, in April 1986, Weinberger repeated that private businesses would be able to utilise technology commercially; but he also warned that 'of course, classified technology would still need to be adequately protected'.

It should be mentioned in this respect that the American armed forces are becoming increasingly dependent on Japan as a source of cheap and reliable electronic hardware. The procurement policy of SDI is to buy the best technology at the lowest cost, and in many cases that may mean buying Japanese. According to Richard Reynolds, director of the Defense Science Office at the Pentagon's Defense Advanced Research Projects Agency; 'Japan has the technological base to become a major supplier to the SDI'. Other people in the defense industry believe that in some key areas, such as optical fibres, chip packaging and opto-electronic connectors, Japanese companies lead American military contractors. Some Japanese doubt this and make the point that if the Pentagon wants Japanese technology and components, then it can buy them on the open market. The suppliers would not violate the ban on the export of weapons, that way, because such items also have civilian uses (Johnstone, 1986).

On 9 September 1986, chief cabinet secretary Masaharu Gotodo announced that the Japanese Government will allow participation in SDI. A major factor in the decision was a report of the third visiting team to the United States. The report recommended participation because it could have 'a very favourable impact on the improvement of Japan's technological standards'.

The United Kingdom

Britain and the United States agreed in October 1985 on a framework accord setting forth the terms for participation by British companies in the United States Strategic Defense Initiative. This document has become the first government agreement linking the United States with one of its allies in the planned research program for the SDI. The agreement is expected to serve as a precedent for other allies who have awaited the outcome of the British–American negotiations before seeking to conclude their own agreements with the United States.

At the hearings mentioned earlier in this chapter Mr. LaFalce, chairman of the subject subcommittee stated with respect to United Kingdom's participation in SDI as follows:

> What I am trying to understand is how much research are we going to do in the United States, how much are we going to do abroad? What are the motives that are leading us to leave the borders of the United States?

Because there is unquestionably some pretty good spin-off from military research, at least that has been the case historically. As a matter of fact, Lieutenant General Abrahamson recently set up an Office of Civil and Educational Applications within the SDI Program so we could obtain the commercial spin-off from the SDI research.

Now, it is going to be much more difficult to obtain commercial spin-off from technological research done abroad. It is also, it seems to me, going to be much more difficult to control the demonstration of that research information that we do not want disseminated if it is done abroad, as opposed to having been done in the United States, and it is also relevant whether the contract was with foreign governments or with foreign firms.

In reply, Mr. Yonas, chief scientist and deputy director of SDIO stated as follows:

To the extent to which our allies have already made investments, for instance, in facilities, innovative concepts, for instance, in the United Kingdom we see investments that already exist which they are bringing to bear, which they are bringing to the table in terms of these cooperative programs, that is a real investment. It is their brain power, their facilities and their prior ideas that they are bringing to the program, and what we are trying to do is to capitalize on that investment to move forward more rapidly together to solve a common problem. (House of Representatives, 1985.)

The possible programs that Britain could be active in were briefly described in the *Financial Times* of February 19, 1986.

Mr Yonas sees "The Ministry of Defense's Royal Signals and Radar Establishment at Malvern Worcestershire is a repository of expertise in infra-red surveillance unrivalled in the U.S. According to Mr. Yonas: infra-red sensing is the key to the surveillance problem. Infra-red is the spectrum of invisible 'heat' waves given off by anything above absolute zero in temperature. Missiles emit different infra-red frequencies at different stages in their flight trajectory, from the hot boost phase through the cool mid-course stage to the hot atmospheric re-entry phase. Infra-red is the eyes and brain of the whole system, right across this trajectory, Dr. Yonas says. Moreover, Malvern has developed a semi conductor technology based on its own discovery of a compound called cadmium mercury telluride (CMT), and married this in a unique manner with silicon chips to process the infra-red signals sensed by CMT. He talks of these sensors as 'little jewels' — the combination has the value of a two or three carat diamond, he says. 'And sensors are the big money in the SDI programme, billions of dollars.' The U.S. team also invited the industrial laboratories poised to exploit Malvern's CMT technology, namely GEC's Hirst Research Centre in Wembley and those of Mullard, the Philips subsidiary. But, in addition, Britain has ideas worth backing concerning new weapons, including speed-of-light weapons that can discriminate between genuine targets and decoys designed to confuse the defenses.

At the Rutherford Appleton Laboratory of the Science and Engineering Research Council near Oxford, the SDI scientists inspected Sprite, a krypton fluoride laser pumped by an electron beam, built as part of a central facility for British universities to do experiments involving large pulses of laser light.

This facility, which has already permitted U.S. scientists to use its lasers under contract to perform experiments, is discussing a 'fairly major' SDI project, says Dr. Mike Key, its director. The SDI sees this kind of laser as a promising way of destroying decoys from the ground, leaving other weapons a clear view of the genuine targets.

At the Culham Laboratory of the U.K. Atomic Energy Authority, also near Oxford, a substantial research contract has already been negotiated with the SDI to develop a beam source for a potential space weapon called the neutral particle beam. (*Financial Times* (1986 op cit.).)

Other British research groups that have been invited by the SDI-office are researchers at Heriott-Watt University in Edinburgh for work on optical processing computers and a research group at Brunel University for their research into laser gyroscopes (to keep space platforms steady as they orbit above the earth).

There is a lot of criticism with respect to costs, complexity and contractual difficulties in dealing with the SDI-organization in Washington. British firms are reported to find it impossible to bid for 'Star Wars' research from their side of the Atlantic. The United Kingdom government recently appointed a top industrial manager with inside knowledge of the United States' defense industry to increase British participation in SDI. This industrial manager will be paid by his own company and will be joined by ten other industrialists on the same financial basis.

Up until July 1986 five contracts worth less than $1 million have been signed by British firms for SDI research. A dozen more are expected. In June 1986 the United Kingdom Defence Secretary George Younger announced two government-to-government contracts worth $14.3 million. Another four contracts are expected, bringing the value to around $30 million (see *New Scientist*, July 1986).

France

An effective space-based defense against missiles cannot be built for at least a quarter-century, and even then could be pierced by the relatively small French nuclear arsenal, a French government commission concluded in May 1986.

The commission, headed by J.F. Depech, director of research at the French National Center for Scientific Research, concluded that neither superpower would be able to base an anti-missile defense in space for at least twenty five years, and that such development would require 'spectacular technological breakthroughs'. The Report stated that 'even in the case of extremely rapid technological progress, accompanied by considerable appropriations, we find that it does not seem very realistic to expect that one of the great powers will deploy a space component sufficiently large to be reasonably effective before the year 2010'. In an article in the *Washington Post* it was stated that French defense officials may draw on the panel's advice as they plan the modernization of their nuclear force. The French government, elected since the report was submitted, is expected to continue the previous government's policy of maintaining an independent nuclear force while keeping a distance from SDI. What impact this report may have on policy is unknown (see Hiatt, 1986).

President Mitterand is opposed to SDI whereas Prime Minister Chirac favours participation. Chirac's Minister of Defense André Giraud is a supporter of SDI. Giraud has the opinion that French industry may have much to gain from it. This opinion is strongly supported by France's electronics companies, especially in defense and aerospace.

The German Federal Republic

The United States and the German Federal Republic signed an agreement regarding German participation in SDI in April 1986. A Cologne newspaper published the agreement together with the exchange of letters between the chief German negotiator for SDI,

Lorenz Schomerus, and Richard Perle from the United States Department of Defense. The publication came the day after the West German parliament discussed the agreement in secret session. Siegfried Mann, head of the Association of German Industry, said the relevation will hurt German interests. General Franz Joseph Schults, former head of NATO's middle-European Strike Force, is reported to have said: 'If we cannot even keep the agreement a secret, the Americans will wonder how we'll do with the other things, and whether cooperation is possible'. Security was a pre-occupation of the agreement. It guarantees that the United States 'insofar as law, national policy and international commitments permit', will apply the same conditions to SDI contracts awarded to German as to American firms. The United States government promises to share information relevant to SDI projects with German companies and to allow them to bid on equal terms for contracts. Both governments promise to take all possible steps 'to prevent transfer of sensitive SDI technology to forbidden destinations'. Classified technical data and computer software developed or exchanged under the agreement are singled out, with each government agreeing that it should be responsible for preventing 'unauthorised publication'. The United States Department of Defense is to give instructions with each contract on what is or is not classified, but the agreement states that 'as a contribution to this desired security, information that one government has given to the other (such as the agreement itself) is designated as secret'.

A sensitive point in the deliberations between the West-German government and the SDI Office was, until the last moment, the status of West Berlin companies. In the final agreement the United States has acceded to the German requests to permit West Berlin to participate in the research. The agreement is supervized by German financial, rather than military, authorities. The agreement has also been signed by the West German finance minister, Martin Bangemann. Chancellor Helmut Kohl has always stressed technological cooperation rather than the military aspects of SDI so as to avoid straining West Germany's relations with the Eastern Bloc. Possible spin-offs were an important argument for West German participation. The agreement consists of two parts, a general agreement on technology transfer, and a memorandum of understanding concerning participation of West German companies in SDI-research. The strong German demand for some guarantee that

German companies will share fully in the technology developed under SDI, has been refused by the United States. There has been substantial public debate in West Germany on the economic benefits of SDI. Criticism there says that SDI-participation is not the way to catch up technologically and that capital and human resources will be withdrawn from civil areas (see Eisbach, Rilling, 1985; Bertram, 1985).

Other countries

In addition to Great Britain, Japan and the Federal Republic of Germany, United States Secretary Weinberger signed a Memorandum of Understanding with Israel on May 5, 1986, and with Italy in September 1986. In a meeting in Lissabon, March 1985, socialist parties of the NATO countries reached agreement to refuse participation in SDI, but the subsequent withdrawal of Italy made the agreement inoperative. The Norwegian government has decided not to participate, and the Danish government might join. The Dutch government will not participate in SDI: 'the costs and efforts won't counterbalance an expected positive effect on Dutch technological capacity and the Dutch economy' according to a Netherlands' government statement. Nevertheless Dutch companies won't be prevented from taking up contracts with the United States government. Company officials of Philips have stated that they are interested in SDI and would like to participate.

SDI is no European economic elixir

We already expressed doubt about the United States' objectives in encouraging foreign participation. At a conference in the United States on International Space Technology, industrial defense-experts remarked that they expected European participation to be less than 1 per cent of the total (Lowndes, 1986). The SDI-office seems only interested in those areas where European companies or research groups are far ahead of United States competitors. It can also be noted that the American record on cooperation in production and research with Europe in the defense area is poor. Project after project has been scuttled by service representatives in the Pentagon, by competing American firms, or by members of Congress whose constituencies house American competitors (Dean, 1985).

The example of the Heriott-Watt University in the United Kingdom shows what other European participants might experience with respect to economic benefits of participating in SDI. The University signed an agreement with SDIO to package and transfer its know-how in optical computing. However, it is not involved in the 'Star Wars' programme itself. Indeed, in the USA there is an SDI consortium of seven universities and seven companies established by the Department of Defense, and it is into this broad-ranging context of research and industrial competence that Heriott-Watt's knowledge is being transferred. Clearly, there is no possibility of the United Kingdom benefitting from the work within that group which has been enabled or accelerated by the inward transfer of United Kingdom technology (Arnold, 1986).

For European countries the strongest positive attraction of the American missile defense program has been the possibility of European participation in an important new technological development from an early stage. But foreign participators will meet a problem, in that their research groups are not at all familiar with the American defense-acquisition culture.

References

'France may ditch Eureka and join Star Wars', *New Scientist*, March 27, 1986.

'Marching to the SDI drumbeat', *New Scientist*, July 10, 1986.

Arnold, E., *Policies for electronics, computing and the information technologies in four smaller OECD economies: Belgium, Canada, the Netherlands and Switzerland*, A report to the National Swedish Board for Technical Development, Brighton SPRU, 1986.

Bertram, C., 'Lockrufe aus Übersee', *Die Zeit*, April 26, 1985.

Dean, J., *SDI & NATO*, papers on Strategic Defense, Cambridge Union of Concerned Scientists, 1985.

Dickson, D., 'Europeans Wary of U.S. Offer on Military R&D', *Science*, Vol. 232, April 18, 1986.

Eisbach, J., & Rilling, R., 'SDI en de beperkte spin-off van militaire R&D', *Tijdschrift voor Politieke Ekonomie*, p. 9, nr. 2, 1985.

Hiatt, F., 'French see "Star Wars" at least 25 years away', *Washington Post*, May 1, 1986.

House of Representatives, *Impact of Strategic Defense Initiative on the U.S. Industrial Base*, Hearing for the Subcommittee on Economic Stabilization, Committee on Banking, Finance and Urban Affairs, December 1985.

Johnston, B., 'Japan postpones launch of its Human Frontier Program, *New Scientist*, May 15, 1986.

Lowndes, J.C., 'European SDI share could amount to 0.1%', *Aviation Week and Space Technology*, January 20, 1986.

Marsh, P., 'Star Wars is "all chips versus nukes" says US expert', *Financial Times*, February 19, 1986, p. 13.

Part II Military or Industrial Technology Policy

4 Government and the development of industrial technology

The benefits currently accepted to be associated with industrial innovation and technical change are such that governments in most industrialized countries feel that they cannot do without explicit technological innovation policies. The pressures to construct such policies have grown as a result of a number of factors, including the worldwide deterioration of the economic climate during the 1970s, changes in the international economic order, balance-of-payment problems, competitiveness of industry, and last but not least, the general acceptance of the case long argued by many economists that innovation and often related technical change play key roles in stimulating economic development.

At the national level innovation policy has become a point of convergence between industrial policy and science and technology policy. Such policies are attracting more attention now because traditional economic policies are not supposed to be able to formulate an effective response to the current economic stagnation. On the other hand traditional science and technology policy is not expected to provide sufficient impulse and direction to technical change. Technical change and economic development however can neither be separated from a larger societal context nor from the micro level, that of the individual firm. The subject of 'technology and change in industrial societies' and the role of government in this process can be treated at three interrelated levels: society at large, the individual firm and the national economy.

Society and change

It was Schumpeter in 'Business Cycles' who first advanced the notion of radical innovations as a major factor in the recurrent crises of structural adjustment. Schumpeter thereby pointed to such important exogenous factors as wars and harvest failures facilitating adjustment in institutions and other structural elements due to rather drastic value changes (Schumpeter, 1939). Since the strategies of the actors involved in the adjustment processes depend heavily on the way in which relations between groups and individuals are organized, the importance has to be recognized of such structural elements as markets, decision-making processes, public services, government administrations, national or transnational corporations, labour unions and other associations, and intergovernmental organizations. It can be assumed that in the course of the next quarter century, these structural elements will undergo substantial change — some will develop, others will have to contend with mounting difficulties and still others will survive by adapting. Some may also be changed or transformed by governments wishing to use them in order to find better answers to long-term problems.

Present institutions often temper the pace of technical change and thereby the adjustment process. These institutions have most times been established in a different era and for different purposes. For example, the traditional role and structure of the PTTs (Post, Telegraph and Telephone) in industrialized countries with respect to telecommunications is not necessarily the most advantageous one from a present day view on the development of the telecommunications and the services industry. The role and structure of centralized electricity generation, distribution and pricing does not necessarily warrant the desired development of decentralized energy systems. The role of the labour unions cannot but differ in a period in which the position of workers-councils is increasing and unemployment figures are high. Decision-making processes have changed and will continue to do so over time; one of the main problems being the level at which decisions are made. On the one hand decisions regarding nuclear energy can hardly be other than centrally taken. With respect to the introduction of modern production technology on the other hand, a decentralized decision-making process seems to be a prerequisite. Although current environmental problems may warrant a stricter regulatory effort, our current insight into the eco-system is insufficient for it to play a role in economic theory.

The same holds true for the relationship between technological development and economic development. Economists largely treat technology as a 'black box'. In a period when the role of technology in society and in the economy is large and still growing, the structure of economic modelling is largely neglecting the specific role of technology with respect to new products and processes. Finally, there is the discussion with respect to the relationship between technology, employment and work. It is likely that the main actors in society will not reach agreement on totally removing the links between work (and employment) and income for some time to come, at least not as a matter of principle. The main actors concerned, however, may strike some bargain in which there may be 'work for all' — including income for all. Even this approach will imply work to be placed in a new institutional framework and new concepts of work and employment.

It can be concluded that technology is a forceful change agent in society, that changes ought to fit values of the population or of the subject social groups and that new structural elements are often required to accommodate the changes concerned. It is the role of many actors involved and especially of the political system, to develop insight into the role of technology and to adjust structures accordingly.

Change and the individual firm

The individual firm is confronted with numerous external and internal changes. Basically the individual firm went through a process of strategic positioning along the lines of the manufacturing of products, the realization of product-market combinations and the operationalization of these product-market combinations within the framework of a social and physical environment. Presently, technology is becoming a strategic factor in the development of the firm. Ansoff's 'Corporate Strategy' was first published in 1965 and pointed at two interrelated elements: diversification and the so-called product-market matrix. Diversification by way of new products serving new markets has not always been a successful approach and the present trend is to concentrate more on the 'core' of the business. This latter development is frequently taking the form of spinning-off the diversified part of the operation. When considering product-market-technology combinations a different matrix is now appearing. Technology in this case is to be considered as a strategic element for the firm; either in the form of production technology or in the form of product quality in a broad sense.

Technology as a strategic element necessitates more attention to access of technology. This in turn often necessitates new and different relationships between universities and industrial firms as well as between industrial firms traditionally operating in identical as well as in different sectors.

The importance of technology in the process of future industrial development makes it essential that appropriate structures be built to effectively operationalize and make benefit of the complementarities that can exist between firms of different nature and size. These complimentarities can be expressed in terms of technology, market-power, finance etc. and are often dynamic. The relative roles that the firms play, both separately and cooperatively, are likely to vary over the industrial cycle: from strategic planning to research and development, from production to marketing. The different time horizons on which the firms operate and their different company cultures often constitute hindrances to effective collaboration. Key problems for the individual firm include 'management of change' both internally and with respect to its position on the market.

Change and the national economy

During the past three decades, at least up to the end of the 1970s, macro-economic policy in the Western world has concentrated mainly on the maintenance of full employment largely through the control of financial flows and through demand management. Post-war economic development up to the mid 1970s can be characterized as an era of continuous economic growth accompanied by some minor fluctuations about the growth path.

It is important to stress the structural character of the present economic situation. In the late 1960s, a large number of post-war industries entered simultaneously the maturity- and market saturation-phase of their life cycles. A dutch study by Geldens showed, that no less than 75 per cent of Dutch industrial value-added was generated in industries in the saturation or declining phase of the industry cycle (Geldens, 1979). In this phase of the cycle keywords are: rationalization and growing automaticity, growing and manufacturing unemployment and price competition. Under these conditions innovation is necessary in order to initiate new products and to enable industries to enter the first phase of a next generation of growth cycles.

The implication of this structural interpretation of the present crisis is that it should not be tackled by traditional demand management policies alone, but by adding a different kind of policy: reindustrialization and technology policy. While an adequate traditional macro-economic stabilization policy is necessary to create a favourable climate for innovation and reindustrialization, it is, by itself not a sufficient condition to induce the necessary innovative activity that constitutes a primary element in economic and social development. Certainly poor macro-economic conditions are not favourable to radical innovation and weak aggregate demand and uncertainties with respect to the future both have their negative repercussions regarding investment behaviour in general, and to investment in new product and process technologies in particular.

With respect to the size of future innovative efforts, it is more than probable that reduced manufacturing profitability and subsequent reduction in real company R&D outlays diminish the sources of future innovations and hence the opportunity for future economic growth. Uncertainty concerning future economic developments may give rise to 'risk-averting' behaviour, consisting of a shift to less radical projects with short payback-periods, thus influencing the nature of the innovative efforts: a tendency towards incremental innovation to the detriment of radical technological change. McCracken ascribed the economic problems of the last decade up to about 1973 to a combination of exogenous disturbances such as the oil crisis, the sharp rise of prices of food and raw materials, the breakdown of the exchange-rate system and the coincidence of elections in a number of major countries in 1972, as well as to some avoidable errors in economic policy. Although we do not altogether agree with this analysis McCracken's statement can be supported to the extent that an important task of government's macro-economic policy should be 'the stabilization of expectations and the rebuilding of confidence' (Zegveld, 1985).

Industrial technology policy

A balanced and integrated industrial technology policy contains three principal and interrelated elements determining innovative potential. The three elements are:

— Technological opportunity.

— Structure of the industrial sector.

— Size and structure of market-demand.

It is apparent that in an approach towards reindustrialization defined as 'the structural transformation of industry into higher added value, more knowledge-intensive sectors and product groups, and the creation of major new technology-based industries serving new markets', science and technology are the necessary, although not sufficient elements.

Technological opportunity

An essential element in industrial technology policy is the transfer of scientific and technological knowledge to industry. 'Knowledge' as an input-factor is transferred to industry in two main ways. The first way is that knowledge is incorporated in the labour supply to industry via the labour market. The currently perceived importance of education and training can well be illustrated and measured by the considerable financial allocations to this area; the financial means in question represent a significant share of overall government budgets. At the same time, relatively little attention is being paid to the development of coupling mechanisms between the educational system and its 'clients', an important client being industry. The second technology transfer path consists of direct knowledge and hardware transfers to industry from the government co-financed technical and scientific infra-structure.

Within the framework of industrial technology policy, it is clearly important to restructure the pattern of allocation of public funds, skilled manpower supply and public R &D facilities in such a way that better match the requirements of the market sector. To achieve this, of course, it is equally clear that industry needs to establish as precisely as possible, its perceived set of current and future needs. Incentives should thus be provided to the scientific and technical infrastructure to promote a stronger orientation of its resources to answering the pressing scientific and technical requirements of the market sector. This stronger orientation to the problems of the market sector may be achieved not only by changes in the methods of funding infrastructural research, but also by establishing new and different relationships between industry and the infrastructure.

Structure of the industrial sector

In a model of post-war industrial evolution the changes in industrial structure can be described as being associated with the maturization of a set of techno/economic considerations. Also the often described role of New Technology Based Firms. (NTBF) in the emergence of the semi-conductor and computer aided design (CAD) industries in the United States presents clear evidence of the availability of rapid diffusion processes. These points suggest that the emergence of new techno/economic combinations is often associated with the innovative and/or funding endeavour of large corporations coupled to vigorous entrepreneurial activities of NTBFs. More recently the emergence of the new wave biotechnology industry — begun by spin-off entrepreneurs from universities — again illustrates the importance of NTBFs during the early phases of industrial evolution. In addition, it demonstrates the increasing small firm/large firm complementarities taking place in the infant biotechnology industry.

These points suggest the important feature of the emergence of new combinations in the system of dynamic complementarity between small (new) and large (established) firms. Thus an important feature of industrial technology policy lies in stimulating the appropriate complementary dynamic between large and small firms.

Complementarity and collaboration can, of course, exist throughout the industry cycle; for example between 'colleague' suppliers operating in the same market, the form of collaboration varying from joint research and development to production and sales. Complementarity may also take the form of contacts between manufacturer and user or between manufacturer and supplier(s). It is clear that these contacts should be stronger the more technically sophisticated and complex the product is involved. Large, conglomerate corporations have demonstrated remarkable corporate flexibility and dynamism during the rapid technological catching-up and subsequent overtaking of Europe in some areas, achieved by Japanese industry. Even here, however, a system of very many 'tied' subcontractors, has provided the industrial giants a great deal of flexibility in certain aspects of their operations.

From the above description it is clear that an industrial structure of the 1980s and 1990s best suited for innovation and technical change is one of networks of firms. It is the structure that has been proposed by the Wagner Committee in the Netherlands in 1981 for activities regarding the so-called Main Priority Areas.

An important element of the structure of the industrial sector is the financing system. Just as governments have begun to provide special funds to support R&D activities of existing small and medium size firms, such as the Innovation Stimulation Regulation (INSTIR) system in the Netherlands, so also have they taken steps to increase the flow of venture capital to stimulate the establishment and growth of NTBFs. Venture capital schemes have taken a number of forms; both wholly public or private and public/private mixes. The traditional risk-avoiding financing institutions, including pension funds, are now getting more involved in the venture capital business. Although the present schemes would still need considerable development, venture capital is now available and adds an important and necessary element to the structure of the industrial sector.

To function effectively, cooperation between firms must, at least partly, rest on social organization. We have mentioned networks in this context. A second aspect of social organization is the system of industrial standardization. There are four basic types of industrial standards:

— Information standards provide terminology and test and measurement methods for evaluating and quantifying product attributes.

— Compatibility standards specify properties that a product should have in order to be compatible with a complementary product or 'system'.

— Variety reduction standards limit the product to having only a certain range of allowable characteristics such as physical dimensions.

— Quality standards ensure an acceptable level of product performance along several dimensions including reliability, durability, efficiency, safety, and environmental impact.

Each of these types of standards reduces cost and uncertainties inherent in market transactions, allows firms to specialize their product portfolios since they can more readily rely on independent suppliers for inputs and complementary products, and thus achieve economies of scale in production and distribution. They also facilitate the entry of new firms into the activity by encouraging the development of independent sources of input and equipment supply, which reduces the costs of entry, and by limiting the technical uncertainty confronting potential buyers.

Size and structure of market demand

Size and structure of the demand side clearly are key elements in determining innovative performance. Consideration of market structures and dynamics is thus an important element in industrial technology policy. United States and Japanese markets adapt easily and help to create a technically sophisticated supply. From the standpoint of achieving international competitiveness, firms in these countries have first to compete successfully in their national markets where often severe levels of internal competition exist. In the effectively fragmented markets in Europe, fierce competition on more than a national basis is often lacking, largely because of the existence of many non-tariff barriers. Even within the EC, many obstacles prevent the functioning of one 'common' market; technical and administrative barriers to trade, and subsidies to industrial and agricultural firms all distort competition. Trade liberalization, including the harmonization of standards within Europe, by eliminating non-tariff barriers, is a major avenue for European industrial technology policy and would provide a sensible framework for achieving greater competitiveness in world markets.

For a large variety of products governments provide substantial markets and are hence in a position to exercise their market-power in influencing the direction of supply towards high value-added and technologically more innovative products. Thus public procurement policy can be considered to be an effective instrument to promote innovation. Procurement policy fits well with our description of the present economic situation. One of the primary aims of industrial technology policy must be the identification, stimulation and diffusion into use of the new technologies on which future economic growth can be based. The more traditional public policies of support, encouragement, experiment and adaptation of new technologies should be complemented by more ambitious long-term, technology-stimulating procurement strategies. It is, however, clear that there is a great deal of tension between the requirements of innovation-oriented procurement strategies and the current policies of many Western countries which are primarily directed at reducing government deficits and are thereby reducing the market role of governments in the non-military sector.

Military R&D and civilian industry

While we have attempted to make the case for the adoption of coherent national industrial technology policies, the question has to be posed whether the industrial structural transformation process can sufficiently advance in the absence of such policies. In the case of the United States, which has long operated on *laissez-faire*, market-driven lines by large corporations working on an international level, the answer would appear to be 'yes'. This contrasts with the experience in Japan where long-term and consistent national policies have contributed to the remarkable industrial transformation of the past thirty-five years. But it is pertinent to ask a further question: has the United States government, in addition to exercising its advantageous postion of political dominance, been involved effectively in informal rather than formal policies towards industrial and technological change? The answer to this question might be a tentative 'yes'. It is certainly a fact that the United States generally led the way in the creation of the new industries of the post-war era, and there exists convincing evidence to suggest that in a number of cases — semiconductors, CAD, computers, aerospace — the US Department of Defense played an important role (Dosi, 1984; Nelson, 1982; Rosenberg, 1982; Bright, 1979). The Department of Defense played its part both on the supply side (R&D funding) and on the demand side (procurement). This has been summarized for the case of semiconductors by Dosi.

On the supply side they involved:

(a) Stimulation towards precisely defined technological directions and areas in which to allocate R &D efforts.
(b) The incentive toward and the direct financial support of the exploration of different possible alternative paths of technical change. Especially when precise 'trajectories' are not yet well defined there is what one could define as the 'burden of the first comer' (together of course, with the rewards for the first comer).

    This burden, made of the attempts of trying and testing a number of possible areas of advance (which is much greater than that required from imitating industries) has been partly undertaken by public institutions (military and space agencies).
(c) The speeding up of technical progress at what we could define as the 'maximum rate' compatible with the existing knowledge, technology, and experience.
(d) The subsidy of expansion of productive capacity to certain target levels considered necessary for national defense requirement.
(e) A push toward standardization of production.
(f) The lowering of entry barriers for new firms which could find a market with low 'cost of access'.

On the demand side public policies resulted in:

(g) A guarantee of a future market for any innovation corresponding to the required technological features.
(h) The expansion of demand, with associated powerful learning effects upon productivity and unit costs.
(i) (Possible) a subsidy element involved in public contracts which helped to cover fixed costs (such as R &D) that would otherwise have fallen upon civilian sales.

The policies of procurement, R &D financing and explicit indication of the required direction of technological advance, together, operated as a widespread and finalized allocative mechanism of both productive and research efforts. More recently, the United States Office of Technology Assessment has been involved in the preparation of a report which raises the question whether the United States government should actively encourage the commercialization of biotechnology, and if so, whether or not this should be achieved through an explicitly stated formal policy (OTA, 1984). This represents a response to the high level strategic involvement by government in biotechnological developments mainly in Japan, but also in some countries in Europe. Whether or not the United States government does become involved in formulating explicit strategies towards the development of key technologies, the R &D funding and procurement activities of the Department of Defense continue at a high level. There is no reason to doubt that, as in the past, this defense-related activity will have an important influence on both the rate and the direction of technological innovation in United States industry.

In addition to creating a considerable source of civilian R &D funding, such activity by the Department of Defense means that the dogmatic issue of direct formal government involvement with industry can be circumvented. Below we document the recent and ongoing contribution of defense-supported R &D towards civilian industry and technology in the United States and contrast with a civil technology policy system in Japan, a country in which defense funding of R &D is minimal.

The R &D programmes of the United States Department of Defense contribute to the country's civilian technology and to its industrial base. In the past Department of Defense spokesmen have generally referred to this contribution in terms of dual technology and ensuring an adequate industrial-production capacity to satisfy military requirements. In recent years there has begun to be reference to civilian applications *per se* and the importance of a competitive advantage over other countries, particularly Japan. Dr. Richard DeLauer, Under Secretary of Defense for Research and Engineering, is reported to have made the following statement to representatives of the semiconductor industry in February 1983; 'The nth-generation development programme is the United States answer to Japan's government-supported fifth-generation-computer programme'. The United States programme to which Dr. DeLauer referred is formally titled 'Strategic Computing and Survivability'. In view of such goals it seems appropriate to compare the Department of Defense in some respects with Japan's MITI, the main disadvantage of such a role for the Department being the inefficiency resulting from the requirements for secrecy and from the primary emphasis on military applications. Defense-oriented

R&D might be deemed to be a deficient means to achieve civilian innovations. The amounts of money of the Department of Defense budget for research, development, test, and evaluation (RDT&E) are, however, so large that a significant civilian industrial impact is to be expected in spite of this major disadvantage. This budget is estimated at $27.8 billion for the fiscal year 1983 and $29.6 billion for fiscal year 1984.

An example of the impact of the Department of Defense on the scientific and technical community, the employment of scientists and engineers resulting from its activities is mentioned in chapter 2 (see p. 32).

There are a number of paths within the Department of Defense leading to new civilian technology and an improved industrial base, namely:

— the support of research and development leading to civilian as well as military applications;

— the support of research and development leading only to military applications but providing experience with advanced equipment of interest from a civilian standpoint;

— the allowance of funds for 'Independent Research and Development' (IR&D) in areas of interest to the Department of Defense as a cost factor in determining overhead rates for contracts;

— the purchase of large quantities of equipment incorporating advanced technology and requiring sophisticated production techniques;

— the support and the encouragement of university research;

— the implementation of an 'Action Plan for the Improvement of Industrial Responsiveness', which includes as its objectives a sufficient supply of skilled labour to meet the needs of industry and the improvement of industrial productivity.

Other activities can be added to this list:

— the obligation for defense contractors to employ Pentagon's standards leading to standardization of equipment and instruments;

— the stimulation of cooperation between the producers of military-goods leading to coordination, allocation of tasks in order to avoid antitrust-regulations;

— the development of new production techniques which increase productivity and decrease production costs;

— the protection of American products against foreign competition.

The overall responsibility for United States Department of Defense research and development lies with the Under-secretary of Defense for Research and Engineering. It is interesting to note that within his organization there is an Assistant Deputy Under-secretary for Production. The latter's responsibilities include productivity and resources. In addition to the armed services, various defense agencies support research and development. The most important group from the standpoint of new technology is the Defense Advanced Research Projects Agency (DARPA). This serves as the cell for initiating new technologies. After a certain stage of development is reached, the responsibility for programmes is usually transferred to one or more of the armed services.

The information presented above points to important indirect and direct influences of defense R &D spending on civilian industrial development. Set against the considerable levels of civilian R &D funded by the Department of Defense are the potential opportunity costs to civilian market-related developments as a result of the vast sums of cash and large volumes of human capital resources devoted to military developments. There is little doubt that defense is today's growth sector in the United States. Indeed, top-level military spokesmen now claim that defense R &D and procurement expenditures are the dynamo of the United States economy; these same spokesmen now openly claim to be consciously fulfilling a MITI-like role in coordinating and funding a significant percentage of total US R &D activity.

However, as we suggested above, there may be significant opportunity costs associated with this high level of military R &D. There is evidence to suggest that the military industrial complex is depriving the civilian market-oriented sector of many of the country's 'top brains' in such crucial areas as computing and electronics. Large amounts of human capital are being 'locked in' to rather esoteric projects that do not have significant or apparent civilian spin-off potential. Finally, defense procurement is featherbedding some suppliers due to inadequate quality control procedures and lack of accountability. In short, it is being suggested that military R &D spending is seriously distorting the broad thrust of American technological development along paths dictated by military requirements and away from the current and future needs of world markets for civilian goods. This contrasts sharply with trends in Japan. On the other hand there is the possibility that military R &D might be establishing some of the technological trajectories of the future. In addition, there will surely be civilian spin-offs and an increase in the accumulation of national expertise in areas of advanced technology and manufacturing know-how (Rothwell and Zegveld, 1985). Whether these benefits will in the case of SDI, outweigh the costs or vice versa is dealt with in the following chapters of this book.

References

Bright, J.R., 'Technological Forecasting as an Influence on Technological Innovation: Past Examples and Future Expectations', in Baker, M.J. (ed.) *Industrial Innovation*, Macmillan, 1979.

Dosi, G. *Technical Change and Industrial Transformation: The Theory of Application to the Semiconductor Industry*, Macmillan, 1984.

Geldens, J., 'Een profielschets van Nederland', in *Orde van Organisatie-adviseurs: De Vennootschap Nederland*, Deventer, 1979.

McCracken, *Towards Full Employment and Price Stability*, Paris, OECD, 1977.

Nelson, R. (ed.), *Government and Technical Progress*, Pergamon Press, 1982.

OTA, *Commercial Biotechnology: An International Analysis*, Office of Technology Assessment, Congress of the United States, Washington D.C., 1984.

Rosenberg, N., *Inside the Blackbox*, Cambridge University Press, 1982.

Rothwell, R., and Zegveld, W., *Reindustrialization and Technology*, Longman, 1985.

Schumpeter, J., *Business Cycles*, McGraw-Hill, New York and London, 1939.

Zegveld, W., PA Technology Lecture, Wassenaar, The Netherlands, 1985.

5 SDI: defense technology or dual-use?

Several authors have agreed that the United States Department of Defense plays an important role in industrial and technological change. In this chapter we focus on the policy character of the SDI program. Is it a defense R &D program, or is it a substitute for an industrial technology program? Does the American system not accept direct government involvement in the strategy of private industry and is the military route a more appreciated and appropriate way for such involvement?

These questions are at the heart of the defense–economy debate in the United States since it was acknowledged that war production was responsible for pulling the nation out of the Great Depression of the 1930s. A sort of unintentional Keynesianism rescued the economy, a variant of Keynes in which the large demand-creating government intervention in the economy was almost wholly military in origin. According to Galbraith, the military system, and especially the Department of Defense became the heart of a new planning system. This planning system — consisting of a web of interlocking interests, public and private — had supplanted the market as the key economic factor in American life. The State, which provides educational facilities, research and development laboratories, national security, and so on, is an essential part of the planning system. And the military, as the dominant force of power in the state, is visibly entrenched in the systems' foreground. The Department of Defense, Galbraith wrote, 'supports the most highly developed planning in the planning system. It provides contracts of long duration, calling for large investment of capital in areas of advanced technology' (Galbraith, 1978).

In Galbraith's depiction, consumer demand, national security and technical development were fine-tuned and modulated by the planning system. Defense procurement was the *sine qua non* of the new industrial state: the biggest, the most amenable to centralized planning, the least accessible to market forces, the most comforting to scientific genius, the most resistant to criticism and scrutiny. Galbraith's part in the defense-economy debate was that he codified the idea that large complex organizations — public and private — were the defining characteristics of the United States economy, and that planning was the dynamic of these organizations. This view was important in establishing the notion that the Department of Defense was an immense economic planning agency with profound effects on the economy as a whole and certain aspects of that economy (regional development, technology) in particular (Tirman, 1984). Tirman in his intelligent historical overview of the defense-economy debate gives the industrial economist Seymour Melman the role of the other side of the debate. According to Melman military spending is not 'good' for the United States economy in its demand-stimulating role; on the contrary, the DoD procurement process damaged the civilian economy in several different ways. He criticizes especially the inefficiency of the military production system (e.g. cost overruns are expected and built into the system by the accepted practice of underbidding to secure a contract) that influences the whole industrial sphere. Inefficiency is created by the higher costs endemic to defense contracting, bidding up of raw materials, services, and salaries; the very weight of bureaucratic rules and paperwork that encumber the private companies engaged in military work.

Melman warns against the Pentagon's influence and dominance over the civilian economy which is apparent in its control over capital, its control over research and development and its control over means of production and of new technical personnel. As a planning system, the military saw to it that it had sufficient funds ($1.5 trillion in 'capital' in 1950–75), technology innovation (one third of all United States scientists; 80 per cent of federal R &D monies), and new blood (sponsorship of relevant curricula in colleges) (Melman, 1974).

Tirman concludes that the 'spin-off' argument is key to understanding the defense economy debate. If, as Melman and others maintain, the defense establishment is a drain on the civilian sector without returning a sufficient number of tangible benefits, then the implication is clear. But if the return on the defense investment is not only security, but steady advances on the technological front, then the Melman argument is far less compelling.

In the late 1970s technology achieved a new significance in

international relations. A decrease in the more traditional forms of military and economic dominance in the industrial countries led these countries to use their hegemony in science and technology to obtain a prominent position in international affairs. One can establish this view in the debate on the European answer to the SDI-initiative of the United States and in particular in the way the European countries have reacted. At the initiative of France, a civilian technology program Eureka was set up.

The possession and control of technological knowledge as intellectual capital is becoming more and more a key to economical as well as political power. The two positions in the defense–economy debate can nowadays be found in the opinion of two critical students of United States government policy. Robert Reich criticizes the Reagan Administration for viewing international relations, including the economic contest with Japan, in any terms other than cold war diplomacy. The United States is sacrificing the Nations' economic future to the short term exigencies of national defense and the parochial demands of domestic producers. According to Reich 'America continues to regard the rest of the world through the foggy lenses of cold war diplomacy rather than through the clear glasses of commercial competition'. The only concerted effort by Washington to meet the Japanese threat in commercial high technology is through veiled attempts at sealing off the American market from the Japanese. In early 1982 the Defense Department and the Federal Communications Commission both urged AT &T not to award a large fibre-optics contract to Fujitsu, even though the Japanese company had offered the lowest bid. AT &T has now announced (and with respectful awareness of the fact that AT &T's faith now rests with several government officials engaged in re-designing communications policy) that the contract, and all future fibre-optic contracts would be awarded only to American firms. Reich writes that international economy is of secondary importance to the Reagan Administration. Foreign policy is the domain of the State and Defense Departments; the Commerce Department plays third fiddle.

According to Reich there is no equivalent of Japan's MITI, which is concerned primarily with the future of the national economy and its place in the changing world economy (Reich, 1984). Tirman comes to an opposite conclusion. The Pentagon's impact on the economy, and on technological advance can no longer be ignored.

Theorists of industrial policy should take a closer look at the Pentagon's industrial policy (Tirman, 1984).

Defense budget dominates the federal budget

The question on the industrial or military character of military R &D and especially of the Strategic Defense Initiative seems rather self-evident regarding the growing part of military R &D in the federal budget. In the President's budget request for fiscal year 1987 military programs dominate the R &D components of the spending plan. This is about 73 per cent of the nearly $61 billion that the Administration intends to obligate for the conduct of R &D and the $1.7 billion for construction of related facilities. This is a growth of $8.8 billion or 16 per cent which stresses Reagan's enthusiasm over research as the elixir of the industrial and military process.

Unlike 1986, defense research activities do not account for quite all of the $8.8 billion increase stated for R &D funding, but they come close. Funding for the Department of Defense's R &D programs would increase 25 per cent, or $8.3 billion to $41.8 billion. The department of Energy's Nuclear Weapons R &D programs would rise 16 per cent, or $363 million to $2.6 billion. The defense budget dominates this 1987 federal budget, and has done so for the past quarter century, falling below 50 per cent of federal R &D obligations only in the single year 1966.

In 1960 defense research activities constituted no less than 80 per cent of the federal R &D funds. In 1958, spurred by the humiliation of Sputnik, President Eisenhower ordered a massive scientific build-up. Congress passed the national Defense Education Act — a four year $887 millions program of student loans, fellowships and other aids and expanded the role of the National Science Foundation. Between 1960 and 1970, the physics faculties of U.S. universities practically doubled in size.

This build up stopped abruptly at the end of the 1960s as the nation struggled with the Vietnam crisis. The defense budget declined sharply and hovered around the 50 per cent level until the early 1980s when it rose swiftly again.

What has emerged since the Second World War in the United States is a system in which the federal government has become the

Table 5.1 Federal funds for R &D by major budget function 1960–84

	Dollars in billions			As percent of total obligations	
Year	Total	Defense	All other	Defense	All other
1960	$ 8	$ 6	2	80	20
1961	9	7	2	77	23
1962	10	7	3	70	30
1963	13	8	5	62	38
1964	14	8	6	55	45
1965	15	7	7	50	50
1966	15	8	8	49	51
1967	17	9	8	52	48
1968	16	8	8	52	48
1969	16	8	7	54	46
1970	15	8	7	52	48
1971	16	8	7	52	48
1972	17	9	8	54	46
1973	17	9	8	54	46
1974	17	9	8	52	48
1975	19	10	9	51	49
1976	21	10	10	50	50
1977	24	12	12	50	50
1978	26	13	13	50	50
1979	28	14	14	50	50
1980	30	15	15	50	50
1981	33	18	15	55	45
1982	36	22	14	61	39
1983 (est.)	38	25	14	66	34
1984 (est.)	46	32	14	70	30

Note: Detail may not add to totals due to rounding. Estimates given for 1984 may change significantly as the result of congressional action on agency budget requests. Data for 1960–77 are shown in obligations; data for 1978–83 are shown in budget authority.

Source: Executive Office of the President, Office of Management and Budget, 'Special Analysis K,' *Budget of the U.S. Government, 1984*, 1983.

dominant purchaser of R &D, but without, at the same time, becoming a primary performer of R &D. The institutional development has thus been the manner in which the federal government has accepted a vastly-broadened financial responsibility for R &D without at the same time arranging for the in-house performance of R &D.

The ratio basic/applied research in the SDI-budget

The question of whether SDI, or in a more general way, the Pentagon is the performer of an industrial or a defense policy depends among others on the height of their budget for fundamental research. If there is substantial civil, commercial spin-off from military projects, then civil and military technological trajectories will more or less converge. The only kind of research that is not directed to applications, civil or military, and which results might be used in both contexts is basic research. So the industrial potential of military R &D programs, among other things, can be derived from the size of the budget for basic research. What are the implications in the case of the defense budget and in particular the SDI budget? Rosenberg found that the defense budget is far more development-intensive than the rest of the federal budget. The result is to impart a strong bias in the overall federal budget in favour of the development component. Given the large share of total R &D expenditure constituted by the federal R &D budget, the result is a strong development bias in the composition of overall federal R &D (Rosenberg, 1985).

The 1982 federal budget, broken down into defense and non-defense components, and the share of basic, applied and development expenditures within each total can be seen in table 5.2. The largest items of the Department of Defense R &D budget involve the development of advanced weapons systems, construction and testing of prototypes, etc. Conversely, the share of the Defense Department's R &D budget that goes to basic and applied research is, by far, the smallest of any major government agency.

A further aspect of this development emphasis in defense R &D is a high degree of concentration of the federal R &D budget in just two industry sectors. In 1981, more than half of federal funding for

Table 5.2 1982 federal R &D expenditures

1982 Federal R &D Expenditures (% share)		
	Defense	Nondefense
Basic	3.2	33.7
Applied	11.0	35.3
Development	85.8	31.0
	100.0	100.0

Source: Congressional Budget Office, 1984

R &D went to aircraft and missiles, and almost a quarter went to electrical machinery. Thus over 75 per cent of total federal R &D went to these two industrial sectors. Non-electrical machines was a distant third and motor vehicles and other transportation equipment fourth (see Table 5.3).

According to Rosenberg military expenditures have clearly resulted in building up the R &D capability of private industry, most directly in the design, testing and manufacture of military hardware. This R &D spending has had certain distinct characteristics: it has been over-whelmingly development, it has been directed toward categories of products for which the federal government was also, simultaneously, conducting large-scale procurement programs, and it has been marked by an extremely high degree of industrial concentration.

The Reagan Administration has given basic research preferential treatment. Measured in obligations total federal support for basic research would rise from $7.78 billion in 1985 to $7.86 billion in 1986. Of this the hard sciences and engineering would go up 7 per cent, which underlines the Administration's thinking that physics, mathematics, chemistry, biology and various engineering disciplines 'provide the foundation for the long-term achievement of national objectives of a strong defense, continued economic growth and an entranced quality of life', as the 1986 budget document puts it. In the 1987 budget obligations for basic research would increase 5 per cent to $8.3 billion.

Slightly less than half the total is destined for academic research, according to John P. McTague, acting director of the Office of Science & Technology Policy. As in 1986 the Administration's budget proposal emphasized basic research in the physical sciences and engineering. Funding for those areas would rise 7 per cent to $4.3 billion. Funding for basic research in life and other sciences would increase 2 per cent to $4.0 billion.

A closer look at basic research, though, indicates that non-military growth would be negligible. Inflation estimated to be 4.3 per cent in 1985, would cut into funding increases. It also happens that the DoD's basic research budget category would account for $971 million, a 16 per cent increase on 1984, when the Pentagon shifted all Star Wars projects from the basic research category to two applied research-budget categories. The higher request for basic research reflects the Pentagon's response to the urgings of Keyworth, the then President's Science Advisor, and DoD's former Under Secretary for Research and Engineering, Richard DeLauer, for support of science in universities. Somewhat more than half of all the DoD basic research money goes to on-campus investigators. The rest is shared more or less equally, depending on service, by profit and non-profit industrial centers (such as Battelle Memorial Institute, General Electric and EG & G Systems) and a variety of off-campus research organizations, national labs and in-house units (ranging from MIT's Lincoln Laboratories, SRI International and the Air Force Geophysics Lab to Lawrence Livermore and National Bureau of Standards (Physics Today, 1985).

The same picture of the general military R &D budget can be sketched for the SDI-budget: industry will receive the bulk of SDI's spending.

The top ten companies doing research for the Strategic Defense Initiative received almost $500 million in 1985, about one third of the budget. This sum will rise as the program gathers speed. Many of the corporations with strong positions in the strategic defense market are the same ones that have developed the United States offensive nuclear arsenal: Lockheed, TRW, Rockwell International, Boeing, AVCO and Martin Marietta. Of the top ten companies, the first nine also receive money from NASA to work on producing a space station. SDI R &D contracts are often not an end in themselves for defense contractors, but are a step towards the potentially much more lucrative production contracts that could follow. Both Secretary Weinberger and Advisor Keyworth have asserted that strategic defense costs may use as large a percentage of the

Table 5.3 Federal funds for R&D by budget functions: 1971–84 (dollars in millions)

Function	1971	1972	1973	1974	1975	1976	1977
Total	$15,542.5	$16.495.9	$16,800.2	$17,410.1	$19,038.8	$20,779.7	$23,542.2
National defense	8,109.9	8,901.6	9,001.9	9,015.8	9,679.3	10,429.7	11,863.8
Health	1,287.8	1,546.7	1,585.0	2,068.6	2,170.2	2,350.6	2,628.5
Energy	555.8	574.0	629.7	759.2	1,363.4	1,648.5	2,561.8
Space research & technology	3,048.0	2,931.8	2,823.9	2,701.8	2,764.0	3,129.9	2,923.7
General science	512.5	625.3	657.6	794.4	813.3	857.7	973.8
Transportation	727.9	558.2	571.5	693.4	634.9	630.5	708.4
Natural resources & environment	415.5	478.5	553.8	516.0	624.3	683.0	753.1
Agriculture	259.0	294.4	308.1	313.1	341.8	382.5	456.7
Education, training, employment & social services	215.4	235.3	290.4	236.4	238.6	254.8	230.1
International affairs	31.9	28.6	28.3	23.8	29.0	42.4	66.3
Veterans benefits & services	62.9	69.1	74.3	84.8	94.8	97.7	107.0
Commerce & housing credit	89.5	49.7	50.2	50.8	64.9	68.7	70.5
Income security	144.9	106.3	106.3	70.9	71.9	48.3	55.2
Administration of justice	10.4	23.4	33.2	34.7	44.3	34.8	29.9
Community & regional development	64.6	65.8	78.4	82.1	92.5	108.5	100.9
General government	6.6	7.6	7.4	9.3	11.7	11.9	12.6

Table 5.3 *(continued)*

Function	1978	1979	1980	1981	1982	1983 (est.)	1984 (est.)
Total	$25,998.9	$28,080.4	$30,016.8	$33,319.3	$35,987.7	$38,386.3	$45,577.0
National defense	12,899.4	13,791.0	14,946.4	18,413.0	22,070.0	24,913.3	31,984.1
Health	2,967.7	3,401.3	3,694.3	3,870.8	3,867.5	4,247.7	4,356.0
Energy	3,134.4	3,461.4	3,603.2	3,501.4	2,920.2	2,533.2	2,217.3
Space research & technology	2,961.5	3,008.5	2,981.1	2,696.0	2,584.2	1,883.3	1,897.1
General science	1,050.2	1,119.1	1,232.6	1,340.0	1,353.3	1,489.6	1,718.0
Transportation	767.5	798.2	887.5	869.5	791.0	893.9	1,090.7
Natural resources & environment	903.9	1,009.6	999.3	1,060.5	942.8	913.2	796.9
Agriculture	501.3	551.6	585.3	658.3	692.7	746.6	748.3
Education, training, employment & social services	345.1	353.5	468.0	298.4	226.9	219.9	223.8
International affairs	57.2	116.8	127.3	160.0	165.0	152.1	160.8
Veterans benefits & services	111.1	122.8	125.8	142.9	139.2	157.6	159.4
Commerce & housing credit	76.7	92.7	102.1	105.5	103.9	106.9	91.1
Income security	67.3	56.8	77.2	42.6	31.6	42.0	43.4
Administration of justice	43.7	46.5	45.1	33.8	30.9	33.4	41.9
Community & regional development	91.9	127.3	119.4	104.3	62.5	47.7	39.2
General government	20.3	23.3	22.0	22.1	6.0	5.9	8.9

Source: National Science Foundation, Federal R &D Funding by Budget Function, Fiscal Years 1981–3, 1982; Executive Office of the President, Office of Management and Budget, Budget of the U.S. Government, 1984, 1983; and unpublished data

DoD budget as is currently accounted for by offensive strategic nuclear forces.

A move towards deployment of a full strategic defense system could play the same sustaining role for these firms in the 1990s that production of a new generation of nuclear weapons is playing in the 1980s.

Of the total SDI budget 75 per cent goes to industry, 20 per cent to Government laboratories and 5 per cent of the overall budget that the Administration plans to spend on SDI research over the next five years has been specifically reserved for 'mission-oriented basic research' in universities and small business firms.

In charge of the SDI's basic science program is James A. Ionson who suggested it to General Abrahamson. The Innovative Science and Technology Office (IST), which was set up in the fall of 1984 and disperses the SDI's basic-science money is not one of the Congressionally mandated program elements for SDI. Its mission, according to James A. Ionson, is to 'pursue . . . highly innovative, high-risk concepts that could have a revolutionary impact on the Strategic Defense Initiative'. Although the office will not limit its awards to university-based scientists and engineers, Ionson hopes to 'tap the reservoir of brilliant minds in the academic community and promote university involvement in the SDI-program'.

Topics the SDIO would like university researchers to investigate are wide ranging. Major areas of interest include ultra- high-speed computing, new space-based power sources, novel laser concepts, new types of optical sensors, microelectronic devices that can operate 'in the hostile space environment for extended periods of time', and various studies of 'how the natural atmospheric environment might affect the performance of strategic defense surveillance sensors and directed energy weapons'.

The prime implementation of the IST-office is to put together consortia of universities and businesses and give them three-year contracts to stimulate the flow of technology ideas. In fact, in January 1986 the IST office had already committed $62 million in long-term contracts to support six research consortia including 29 universities. These academic–industrial teams have been set up under the IST's Research Thrust Division to 'address specific science and technology areas that are known to be of critical importance to the success of SDI'. The consortia formed so far will investigate the following areas:

— Non-nuclear space power: a $19 million four-year program involving Auburn, the State University of New York at Buffalo, Polytechnic Institute of New York, Texas Tech, and the University of Texas at Arlington;

— Optical computing: a $9 million three-year consortium consisting of Carnegie-Mellon, Caltech, Georgia Tech, Stanford University, MIT's Lincoln Labs, Battelle Columbus Labs, the University of Alabama at Huntsville, and the U.S. Naval Ocean Systems Center;

— Electronic circuits for computing, sensing, power generation, and directed-energy beams: a $4 million three-year program involving the University of California at Berkeley, Stanford, Purdue, the University of Florida at Gainsville, and the University of Southern California;

— High-speed electronic systems, new dielectric materials for storing power for directed-energy and kinetic-energy systems, and advance power sources for lasers and particle beams: a $2.5 million three-year program undertaken by SUNY at Buffalo, the Naval Research Laboratory, and General Electric;

— Composite materials development: a $15 million three-year program involving research at MIT, Pennsylvania State, Colorado School of Mines, Johns Hopkins, Texas A &M, Brown, Rensselaer Polytechnic Institute, Drexel, SRI International, and the Naval Research Lab. They are to team up with such firms as United Technologies, Martin Marietta, Fiber Materials, Aeronautical Research Corp. of Princeton, and Ultrasystems Inc.;

— Chemical-laser exhaust problems, spacecraft radiation, electromagnetic waves, and particle beams: a $12.5 million three-year program involving investigators at Johns Hopkins's Applied Research Lab, the universities of Arizona, Maryland, Michigan, Iowa, Kansas, California at Berkeley, Utah State, UCLA, MIT, Stanford, NYU, the Naval Research Lab, and the Air Force Geophysics Lab.

Other long-term research consortia to investigate nuclear space-power, propellants, and ultra-shortwave lasers are expected.

Pentagon's influence on university research

Although the percentage of SDI-spending on 'basic' research is rather low, which might affirm the statement that SDI is a military program, on the other hand the growing Pentagon's share of federal funding for university fundamental research may lead to the opposite conclusion.

Since 1980, Department of Defense funding for university research has jumped from $495 million to $930 million, an 89 per cent increase. The next fastest growing source of federal support for university research from NSF grew by 51 per cent over the same period. If DoD funding to off-campus affiliates like MIT's Lincoln Laboratories is taken into account, the Pentagon is now outspending NSF on university research. Campus funding from other major federal agencies — including the Departments of Agriculture, Energy, and Health and Human Services — has grown less than half as fast as DoD support since 1980. Today, DoD research accounts for 16 per cent of all federal spending for university research, up from 10 per cent in 1980. United States' universities are now as strongly dependent on the DoD for research funding as they were in 1968, at the height of the Vietnam War.

This growing dependence on Pentagon support is particularly extreme in certain key fields. In fiscal year 1985, more than 30 per cent of all federal funding for university oceanography research is from DoD, as is more than one half of federally funded university work in mathematics and computer sciences. All forms of engineering research at universities now receive more than 37 per cent of their federal support from the Pentagon and the specialties of astronautical engineering (82 per cent), electrical engineering (56 per cent), aeronautical engineering (54 per cent) and metallurgy and materials engineering (48 per cent) lean most heavily on the DoD (see Table 5.4).

The SDI Organization's Innovative Science and Technology program will accelerate this trend of university dependence on military contracts. While the FY 1985 allocation for SDI innovative programs is only $28 million, this budget is scheduled for a fourfold increase to $100 million in FY 1986. Much of this new money will go to university researchers. Since the IST (Innovative Science and Technology) program is due to receive five per cent of

Table 5.4 The Pentagon's share of federal funding for university research by major field of science, FY 1985 (dollars in thousands)

Field of Science	Total Federal Funding	DoD Funding	DoD Share of Total Federal Funding to the Field %
Mathematics/Computer Sciences	232,065	117,427	50.6
Mathematics	96,059	33,605	35.0
Computer Sciences	93,453	43,136	46.2
Math & computer sciences not elsewhere classified	42,553	40,686	95.6
Engineering	510,435	193,930	38.0
Aeronautical	47,372	25,424	53.7
Astronautical	17,458	14,384	81.0
Chemical	32,693	1,667	5.1
Civil	43,870	4,457	10.2
Electrical	113,810	64,731	56.9
Mechanical	60,101	26,465	44.0
Metallurgy & Materials	115,207	55,237	47.9
Other Engineering	79,624	1,565	2.0
Environmental Sciences	361,092	80,620	22.3
Atmospheric	98,427	16,054	16.3
Geological	94,408	14,572	15.4
Oceanography	158,540	49,427	31.2
Psychology	125,509	24,951	19.9
Physical Sciences	747,999	88,412	11.8
Astronomy	68,338	4,830	7.1
Chemistry	250,532	37,090	14.8
Physics	414,652	43,084	10.4
Other Physical Sciences	14,477	3,408	23.5
Life Sciences	2,873,779	64,423	2.2
Social Sciences	128,408	79	0.06

Source: National Science Foundation, Federal Obligations for Research to Universities and Colleges by Agency and Detailed Field of Science: Fiscal Years 1973–85.

all SDI funding in each of the next five years, it could be a $300 million program by 1988.

Although SDI funds are rapidly pouring into university research departments, the methods used by the SDIO and the goals of the SDI program have stirred a level of controversy at the United

States campuses not seen since the anti-Vietnam war protests. Fear that the secrecy surrounding the program will clamp down on academic freedom to publish and disseminate research findings has dampened the interest of some university scientists and administrators. Most major research universities nationwide have maintained policies from the Vietnam war era prohibiting classified research on campuses. Many wonder whether a politically sensitive program such as SDI will ultimately classify research on campus.

SDI funding is in a specific R&D category that is normally classified because it is used for weapons systems. However, a memo issued by the SDIO in August 1985, and a White House policy statement made in September 1985, both state that SDI university research will be treated as basic research and will not necessarily be classified. Nevertheless, neither document rules out the possibility that the research will ultimately be classified or subject to any other restrictions.

SDIO/IST officials also suggest that professors in charge of SDI research, and possibly even some graduate students, may have to obtain security clearances. This will give them access to all relevant sources of information, including studies from industrial and government laboratories doing classified research. Such a requirement, of course, could prevent foreign students — who at MIT, for example, make up about one-third of the graduate student body — from participating.

SDI research will be carried out primarily by consortia of universities and industrial and government laboratories. So how can research done on university campuses be separated from what is carried out elsewhere in order to keep it unclassified? SDIO officials have themselves commented that it may become necessary to later classify initially unclassified research. Since the innovative science program is essentially a 'mission oriented basic research' program, it is only a matter of time before unclassified research at universities becomes 'critical' for the SDI program. As one congressional aide told *Defense Week*, 'Many Pentagon (research and development) programs begin this way with unclassified efforts. It is an effort to get their nose under the tent'. Controversy over the long-term status of SDI research on campus is prompting some universities to consider their long-standing policy banning classified research. While some universities facing higher political costs will ban SDI research before it becomes classified, others may revise policy guidelines so that badly needed research from SDI will be allowed on campus.

University administrations are already confronting serious public relations problems over accepting SDI research contracts. As Hartung and Nimroody of the Council on Economic Priorities write in their report, 'Pentagon Invades Academia':

> U.S. universities have a strong record in extending the boundaries of scientific knowledge through the conduct of applied and basic research. In recent history, universities have benefitted from a diversified base of federal support — civilian and military — and an open environment in which research may be disseminated freely among scientific colleagues. The phenomenal growth of DoD university funding in general, and the SDI innovative science program in particular, threatens the diversity of funding sources and the open research process — essential strengths of scientific research at U.S. universities.

The DoD's increased share of federal funding to universities has already siphoned off funds that might otherwise have gone to civilian research projects. While the Pentagon's 16 per cent share of total federal funding for university research may seem too small a figure to provide DoD a dominant role in shaping the conduct of university research, several factors magnify its influence. Pentagon funding is far and away the fastest growing source of money for university research. Between 1980 and 1985 DoD support for university work grew nearly twice as fast as any other source of federal money for university research. It is, therefore an obvious target for academic scientists looking for new sources of support.

In addition, the Pentagon is not the only sponsor of military research conducted at universities. Both the Department of Energy (DoE) and NASA sponsor military projects, including DoE's new $3 to $5 million per year SDI innovative concepts program. Finally, the Pentagon has a particularly large share of the available federal funding for research in fields like computer science, electrical engineering, and metallurgy. All these fields are key to the future development of United States high technology industries.

Universities have long been a major source of unbiased basic scientific research, driven neither by military requirements nor by imperatives of commercial product development. Growth of military funding under the Reagan Administration has shifted the balance away from basic research and towards weapons projects.

According to Hartung and Nimroody this trend appears to be worsening: in the proposed FY 1986 budget, DoD research and development is scheduled to increase by 21 per cent, while basic research will grow by only 1 per cent. Within the basic research budget, Pentagon sponsored projects are proposed for a 16 per-cent increase, more than twice the 7 per cent growth rate of National Science Foundation-sponsored basic research projects. Outside the physical sciences, basic research funding will be decreased by five per cent by the next fiscal year (Kistiakowsky, 1986).

The SDI office does not take this view. In response to a query of *Chemical Engineering News*, the SDI office said:

SDI's Innovative Science and Technology program spent only $14 million at campuses in 1985. For fiscal 1986 we anticipate spending about $50 million at campuses, an average of less than $1 million per participating campus. This represents less than 1 per cent of total government funding at universities. All IST research is open to campuses. Also, IST has over 30 research topics, which represent a wide diversity' (Ember, 1986).

Defense or industrial policy

Focussing again on our original question whether U.S. defense policy and in particular SDI is a substitute for industrial policy Gerd Junne, Professor in International Industrial Relations at the University of Amsterdam attempts to understand the dramatic increase in United States defense-spending. It is impossible to interpret this increase as a reaction on changed East–West relations and he states that also West–West relations have given an important stimulation for the United States defense-spending.

As a result of the economic recession, almost all industrialized countries have tried to stimulate their own economy by means of a strong industrial policy. In the United States this has not proven possible, although enough ideas and proposals have been made. The Reagan Administration advocated, very vigorously, a reduction of government influence on the economy. The relatively strong position of the United States on the international market did, however, decline in the 1970s and particularly in the crucial high-tech sectors, where the United States economy had a dominant position in the past. Junne states that the Department of

Defense is now used as a route to pass large amounts of money to strategic parts of industry, to be compared with policies in other countries, to afford these industries to stay ahead in the international re-industrialization contest.

In a society where direct government-interference is regarded as more or less suspicious it is hard to gain enough support for an active industrial policy. Representatives of the 'high-tech' segments of industry would only keep very aloof to such policy. Because of the fact that the 'old' sectors (steel, automobile industry) have more political power than the 'new' sectors (thanks to more employment, more organized labours, more lobbying experience), the funds which could be created in the framework of a more active government policy, would go, for the greater part, to traditional industry.

According to Junne it is therefore more lucrative for the high-tech industries to get more government procurement and R &D contracts via the route of an increased budget of the defense and space programs. Strong opposition against this strategy was not to be expected, because other branches of industry, like the steel and automobile industry derive benefits from military-spending, although presumably less so than the high-tech sectors. This, according to Junne, is why more parts of the United States society could cope with growing military spending rather than with a more specific industry-policy (Junne, 1984).

Only with reference to the threatened national security can large amounts of money be delivered, which otherwise never could be passed on to industry. The difference in research priorities between America with its defense bias and Japan with its market bias is, as might be expected, weaker than the raw statistics might suggest. The makers of science policy in most industrial countries are investing in the same group of core technologies — computers, materials and biotechnology. A review of science and technology policy by the OECD notes that, biotechnology apart, the Pentagon and Japan's Ministry of International Trade and Industry (MITI) are putting their money into very similar kinds of R &D.

In computer science, for example, both are trying to build a 'fifth-generation' computer that can give a rudimentary imitation of human thinking. MITI has underwritten about a third of the development costs of 'very-large-scale-integrated circuits (VLSIC)' the Pentagon has a $300 million development programme in the same area. MITI has a $30 million R&D program on fibre optics; the Pentagon is spending $40 million a year on similar research. Both are also investing heavily in research on new materials such as polymers and metal-matrix composites. Both are spending about $200 million on manufacturing technology, including robots and factory automation (OECD, 1985). The control of the Pentagon on the above civil research and technology subject areas is increasing rapidly. This is reflected in the growing list of military-critical technologies and the Military Emergent Technology Awareness List (METAL).

DoD is working on this METAL-list to identify future technologies that might have military significance. They serve as a lead for the planning of own investments in R&D and generate future control over these new technologies. The identification takes place in the stage of 'mission-oriented-basic' -research before the private sector has picked up the technology and before the technical data are published. If the technologies are identified, DoD will invest heavily in them by R&D funding. In 1979 DoD defined a list of eighteen central technologies as critical to the military: computer-technology, system-technology for big computers, software technology, technology for automatic-real-time control, material-processing and production technology, directed energy technology, LSI-VLSI-design- and production-technology (semiconductors and electronic components), military-instruments technology, telecommunication technology, direction- and control-technologies, microwaves technology, paint-technology for military vehicles, optical sensor technology, technology for deep-sea systems, chemical technology and nuclear technology.

The technologies mentioned in the METAL-list are not taken up in the list of military-critical technologies, but the technologies of the METAL-list are subjected to all DoD-measures. We see a similar kind of development in the Netherlands where the Ministry of Defense recently published a white paper on Defense Technology. In the white paper it is implicitly stated that there are no clear military-technological or civil-technological-trajectories. The definition of the defense–technology concept in the Dutch paper is primarily focussed on the need for technologies relevant for the Ministry of Defense but developed by the private sector. Of the technologies relevant to be included in the defense–technology concept their military applications do not force themselves.

According to the Dutch paper there will be an increasing interaction between military- and civil-technology development during the next decade. This makes civil-technology development as equally interesting as military-technology development (Ministerie van Defensie, 1986).

A strong example of DoD's strategy in industrial technology can be seen in the field of microelectronics. The lion's share of direct federal support for microelectronics R &D comes from the Department of Defense, and is therefore driven by military requirements. In general, DoD-sponsored basic research serves both military and commercial goals. Development activities for military microelectronics, in contrast, do not overlap completely with activities for commercial needs. For example, low-cost, high-volume production capabilities are a high priority for the commercial manufacture of integrated circuits, but the major DoD program to advance IC technology, the Very High Speed Integrated Circuit (VHSIC) program, focuses on design and fabrication of a small volume of highly specialized ICs for use in military systems. DoD support was crucial to early successes in microelectronics. For example, military requirements for smaller circuits were the primary force behind the development of the first integrated circuits. While DoD has continued to fund microelectronics R &D extensively and remains the largest source of Federal support for microelectronics R &D, support from private companies has assumed a much larger role, and includes an entire industry that has sprung up in the last few decades. Today, military applications constitute approximately 10 per cent of the total sales of microelectronic products. Microelectronics technology is crucial to DoD, and its importance will grow further in the foreseeable future. According to two experts: 'it is safe to say that there is not a single western military system that is not critically dependent for its operation on semiconductor integrated circuits.' Other microelectronic devices, such as sensors, are also becoming increasingly important in military technology.

The defense community, like other users of microelectronics, requires high-speed integrated circuits that consume little power. Since United States defense policy has long been based on technological superiority, DoD needs the best possible signal processing and sensor capabilities. Military end users have specific needs as well, including products like ICs that are immune to damage from radiation.

Each of the three services — the Navy, Air Force, and Army — supports activities in microelectronics R &D in DoD laboratories, such as the Naval Research Laboratory, the Air Force Avionics Laboratory at Wright-Patterson, and the Army Electronic Device and Technology Laboratory at Fort Monmouth, and through research agencies. Apart from the services, DoD's Defense Advanced Research Projects Agency is heavily involved in this area. In the recent past DARPA has played an important role in the development of computer 'time-sharing' systems, new computer languages, artificial intelligence and new generations of computers. In cases where the risk is not that great, but the expenditures reach a volume that individual companies cannot spend on their own, the Pentagon often takes the role as financier.

Because of the fact that the Pentagon pays for its own acquisitions on a cost-plus basis (cost of production plus an agreed part of the profit), the Department has a direct interest in keeping production costs as low as possible. That is why the Pentagon finances a Manufacturing Technology Program ($200 million in 1984). A famous result of this program is the numerical-controlled-production machines (now sold by the Japanese industry). Other results are the development of a flexible automation system, CAD/CAM-system. The Pentagon also finances the application of existing production technologies in the Industrial Modernization Incentive Program. In chapter 4 we mentioned the several paths the United States DoD follows to stimulate civilian technology and to improve the industrial base. DARPA is in this respect the most important group. Certain aspects of microelectronics R &D are also covered under a few DoD special programs, such as the VHSIC programs and SDI. The DoD style for funding microelectronics R &D, characterized by long-term investment in R &D with a clear connection to end uses, appears to have been highly successful for achieving military goals. Some members of the microelectronics community would, therefore, like to see the Federal Government aim a similar level of support at commercial needs — a point of view that has gained momentum in the face of the current pressures on the industry. They cite the influential Japanese Ministry of international Trade and Industry (MITI) as a useful model. But opinions in the microelectronics community diverge sharply on this topic. Opponents of further direct Federal involvement in microelectronics R &D for the industry argue that the results of R &D will meet commercial needs and will be available to industry only if the commercial sector directs and carries out the work itself. They view the MITI model as unacceptable in the American context, given the vast differences in industry–government relationships between Japan and the United States.

One example of a plan to start to bridge this difference of opinion comes from the Semiconductor Research Corp. (SRC), a cooperative R&D organization directed by microelectronics industry leaders. SRC currently supports R&D at universities with funds from its member companies. In the next few years, it may ask the Federal Government to match this level of support. The doubled budget would be administered solely by SRC, which would soothe at least the concerns of SRC's member companies about the selection of research topics and the availability of results.

The case of GaAs digital integrated circuits is another example of Defense influence in microelectronics. The microelectronics community had conducted relatively little research on integrated circuits made from gallium arsenide before 1982 when DARPA decided, based on the results of six years of GaAs basic research that they had sponsored, to fund a series of pilot production lines to demonstrate the feasibility of using GaAs for ICs. This announcement kindled the interest of a variety of organizations, and several defense electronics firms mobilized to build a base of activities in GaAs so that they could be involved with the pilot lines. Three new commercial ventures have already spun off from this work. Perhaps even more significantly, there is some evidence that DARPA's interest in GaAs ICs may have helped to convince research organizations such as AT&T Bell Laboratories that the field deserves an intensive research effort. At about the same time, IBM turned its focus for alternative chip technology from Josephson junctions to GaAs. In part because of DARPA's initiative, the activities in GaAs ICs grew in a few years from a handful of isolated efforts in individual laboratories to large programs sponsored by Federal agencies, industry, and universities (OTA, 1986).

Technology transfer problems

Although we have given evidence for the assertion that American defense policy may be regarded as a substitute for an explicit industrial technology policy, there are arguments that weaken this position. In a Background Paper on Micro-electronics Research and Development the U.S. Office of Technology Assessment formulated this dilemma very pointedly: 'keeping information about new defense technologies within the United States is a critical concern for military security; free and wide exchange of ideas and results is an equally crucial ingredient of scientific research'. We mentioned this issue as we discussed SDI funding for basic research. Although Gerold Yonas, chief scientific advisor to the SDI at one time promised scientists that the SDI organization would not classify work or restrict publication under IST program because it was basic science. We also mentioned that all SDI contractors, including those at universities are subject to restrictions at the discretion of the Defense Department. This is caused by the fact that all SDI contracts fall under the DoD accounting category, corresponding not to research but to 'advanced technology development'. They are thus unprotected by a DoD policy established in 1984 by former DoD Under Secretary of Research and Engineering Richard D. DeLauer. The latter policy exempted contracts which corresponded to 'research' and 'exploratory development' respectively. Where this leads to is shown by the following two examples that have been published recently. According to the *New Scientist* of July 10, 1986, the SDI Organisation, which runs the United States, 'Star Wars' program, prevented one of the country's leading physicists from publishing the results of a breakthrough for more than a year. Andrew Sessler, who leads a team at the Lawrence Livermore Laboratory, California, was told he could not publish details of his work on free-electron lasers because it would benefit the Soviet Union. Sessler was asked to withhold publication from March 1985 until May 1986. The SDI Organisation had made a small grant towards staff salaries and the cost of running experiments on a linear accelerator at the laboratory, under its Innovative Science and Technology (IST) program. This gave the organization the right to stop publication of any results.

Sessler, formerly director of the Lawrence Berkeley laboratory at the University of California, proved a theory long held by laser scientists that it is possible to produce very high-energy lasers with a relatively small amount of energy to begin with. In this experiment, a high-powered magnetic field, generated by a magnetron consuming 30 kilowatts, creates charged particles which in turn bombard a material to free electrons. Sessler eventually won the right to publish his results after energetic lobbying by the Department of Energy (which funds the Lawrence Livermore Laboratory) and by senior high-energy physicists. These included Wolfgang Panofsky, a former director of the Stanford Linear Accelerator Centre, where Sessler has also worked. 'It is a case of asking where the rules are, the boundaries between [what is] open and closed', says Panofsky. 'They [Sessler's team] argued, with great merit in my view, that it didn't make any sense to keep this on the dark side of the line'.

The ban was as much of a surprise to the Department of Energy as it was to Sessler. Between 1982 and 1985, the department invested heavily (10 times as much as the IST) in Sessler's work, as part of its research into nuclear fusion as a peaceful and safe means of generating power. When the department ran out of funds to support the work, the IST supplied the balance (Walton, 1986).

The other example is the conference 'Lasers 85' in Las Vegas. Much of the SDI-related material on lasers could only be discussed there in general, since specific data for many advanced systems are classified. For instance, discussing work on advanced excimer devices, Dr. Leroy Wilson of the U.S. Air Force Weapons Laboratory reported at the conference that excellent progress has been made. Systems have been demonstrated, he said, with high efficiencies, good beam quality, and high repetition rates so that high average powers were obtained, but he could not reveal specific numbers for those parameters. According to an article in *Research and Development* the question of continued restriction of laser research data was taken up by several influential speakers at the meeting. During a panel session chaired by Prof S.S. Penner of the University of California — San Diego, Dr. Edward Teller, a member of the White House Science Council, argued in favour of complete openness of SDI research data. For one thing, he suggested, dropping classification of SDI laser R &D would be in line with President Reagan's statement that sharing such technology could reduce tensions between the U.S. and USSR by allaying Russian fears of the unknown. Besides that, he pointed to the great strides made in computer technology where little classification has existed, stating his belief that easy access to SDI data would spur equally fast advances.

However, in the keynote address on United States policy concerning science and technology, Dr. George Keyworth II, who was still the White House science advisor at the time, outlined a different approach. Keyworth interpreted Reagan's offer to share SDI technology as meaning shared control of SDI if it reaches the deployment stage, not as meaning free access to detailed information. He indicated that the concept of shared control would remove Russian worries that might lead them to generate more sophisticated first-strike systems without the need for revealing United States technological advances.

Given the sensitivity of strategic defense efforts and the concern of the present Administration about technological piracy by the Soviets, the security surrounding SDI could be tighter than most defense projects. For example, problems have already developed regarding the commercial uses of VHSICs. Defense Department restrictions on public disclosure of information about the chips is so tight that 'close-up, head-on photographs of VHSIC chips cannot be published because of concerns that the Soviets might be able to determine the chip architecture and reverse engineer the device from such data'.

It is certainly true the Soviet Union might gain access to the technology if less restraint was imposed. Secrecy, however, hinders commercial production of technologies developed by the military. VHSIC chip fabricators, for instance, may have to obtain security clearances for every employee, visitor, and some suppliers. Moreover, chip designs used by the military may have to be altered for use in civilian products. This prompted one manufacturer to ask: 'How much must I change my VHSIC gate-array design to escape security restrictions?'

Publication of scientific results, and participation in scientific conferences are subject to restrictions under the Export Control Act, the Freedom of Information Act, and various administrative orders. So even if the research is unclassified, the SDIO may impose restrictions on who can participate and on what may be published in the open literature and discussed at open meetings. Scientists who want to participate in international scientific conferences where 'military-critical'-technical data are presented and discussed should have an export license. The same accounts for visits abroad, publications in international journals which are available to foreign researchers. DoD and the SDI office have indeed imposed such restrictions in a number of cases. More important than the security aspects in these classification-cases seems to be role of technology in international trade. Because of the growing importance of technology in the economic and industrial adaptation process and the dominant position of the United States as supplier and exporter of technology (with a surplus on the annual international licence-balance of some $6 billion) it seems evident to regard American policy in this respect.

Restriction of scientific and technological communication

Since approximately 1976, the United States Government has pursued an increasingly restrictive policy regarding the flow of scientific and technical information to other countries. The trend to greater restrictions has accelerated recently. The main reason given by the Federal Government for the imposition of a more restrictive policy is concern regarding the extent to which the U.S.S.R. has acquired technology from the United States and its restrictions, which have been imposed in an attempt to influence foreign action (e.g., the curtailing of high-technology exports to the U.S.S.R. following the Soviet invasion of Afghanistan and the ban on exports of the U.S.S.R.–West European gas pipeline). The extent to which other considerations, e.g., a desire to avoid contributions to Soviet economic development, have influenced the policy of more restrictions is not known.

As pointed out in the report *Scientific Communication and National Security*, issued on September 30, 1982, by the National Academy of Sciences, there are five main methods by which the Federal Government can restrict scientific (and technological) communication and which are listed here below.

(i) Classification of information

In 1977 three inventors of a voice scrambler were notified, following submission of a patent application, that disclosure of information on the invention to unauthorized individuals could result in a fine and imprisonment. As a result of protests from the inventors and organizations interested in the free exchange of scientific and technical information when there is no danger to national security, the order was rescinded approximately a year later. It should be noted that with the exception of nuclear-energy information and inventions for which patent applications are filed, the Federal Government has a legal basis for classification only when it has some pre-existing connection with the information to be classified.

At the request of the Department of Defense (DoD) 170 papers prepared for presentation at the August 1982 meeting of the Society of Photo-Optical Instrumentation Engineers in San Diego were withdrawn shortly before the meeting. It appears likely that one reason for this action was the belief by DoD that some of the material should have been classified.

(ii) Export controls

In the case of the withdrawal of papers prepared for the San Diego meeting just cited, the matter of the availability of unclassified information considered to have potential military applications in a form which could be exported also played a role. The International Traffic in Arms Regulations may be used to prevent the communication or the publication of such information.

Other examples which have been widely publicized were the ban on high-technology exports to the U.S.S.R. early in 1980 following the Soviet invasion of Afghanistan and the 1982 ban on exports of goods and technology for the natural-gas pipeline from the U.S.S.R. to Western Europe.

(iii) Contractual provisions

Government contracts for industrial R&D frequently contain provisions requiring the concurrence of the contracting officer prior to the publication of results.

A different kind of contractual provision controlling the export of data is represented by the restrictions on foreign access to some portions of Lockheed's computerized data base *Dialogue Retrieval Service*.

(iv) Voluntary arrangements

A voluntary agreement exists between the National Security Agency (NSA) and the Public Cryptology Study Group established by the American Council on Education. Any paper relating to cryptology is submitted simultaneously to a pre-publication-review group and to a scientific journal. If NSA desires changes and is unable to negotiate them with the author, there are provisions for submission to a review panel. This agreement has been implemented on a trial basis.

(v) Controls on foreign visitors

In February 1980 the Department of Commerce forced the American Vacuum Society (AVS) to withdraw invitations to Soviet-Block scientists a few days before they were to attend a conference on magnetic-bubble-memory devices. This event is also an example of export control because AVS was informed that some of the data to be presented at the conference would fall under Export Administration Regulations and that the 'oral exchange of information in the U.S. with foreign nationals constitutes export of technical data'.

In the same month the Department of State refused to issue visas to eight Soviet scientists who wished to attend a conference on lasers and electro-optical systems and a topical meeting on inertial-confinement fusion. A Russian post-doctoral fellow at the University of Texas who was co-author of a paper at the conference was also denied permission to attend.

In 1981 Stanford University received a letter from the Department of State suggesting that information provided to visiting Chinese scientists in the computer field exclude information on the results of applied research and indicating that the Department of State should be notified before any visits to industrial plants or research laboratories were arranged.

In January 1983 the Institute of Electrical and Electronic Engineers sponsored a Workshop in Micrometer and Submicrometer Lithography at Indian Wells, California. The committee organizing the workshop was informed by the Department of Defense (DoD) that some sessions on electron-beam lithography would have to be limited to United States citizens. Following negotiations on the matter, DoD decided not to impose the restriction.

As part of the Continuing Education Program of UCLA, the Engineering Extension-Short Course Department recently sent out brochures for a course on Metal-Matrix Materials to be held at the University of Maryland in February 1983. The brochure lists as one of the qualifications, proof of United States citizenship. This stipulation was demanded by the course co-sponsor, the United States Department of Defense Metal Matrix Composites Information Analysis Center. The Center is funding the course and providing state-of-the-art applied technical information for it.

Limited distribution as decided by individual agencies

A few restrictions have been imposed on the Lockheed data base (*Dialogue Retrieval Service*) by government agencies. The Department of Commerce, for obvious reasons in connection with its export–promotion program, will not permit foreign access to the United States Export File and the Trade Opportunities Program in the data base. Medline of the National Library of Medicine is only available to Dialogue subscribers in other countries when there is assurance that the library will receive similar data generated in those countries.

Regulation to restrict technology transfer

The primary legal basis for Federal–Government restrictions on the flow of scientific and technical information to other countries is provided by the Arms Export Control Act of 1976, which largely superseded the Mutual Security Act of 1954; the Export Administration Act of 1979; and, in the field of nuclear energy, the Atomic Energy Act of 1954, as amended, and the Nuclear Non-Proliferation Act of 1978. The provisions of the non-proliferation act have been widely publicized, and it will not be dealt with further in this chapter. There is also a law permitting the assignment of a secret classification to patent applications even though the work was performed privately. In the rest of this chapter the legislation and the regulations restricting the export of scientific and technical information will be briefly discussed, as will the role of the Coordinating Committee for Multilateral Export Controls.

The Arms Export Control Act of 1976 and the International Traffic in Arms Regulations

The International Traffic in Arms Regulations (ITAR), which are administered by the offices of East–West Trade and Munitions Control of the Department of State and enforced by the United

States Customs Service, were promulgated under the authority of the Arms Export Control Act of 1976. The regulations, which contain licensing requirements, apply not only to the export of military hardware but also to the export of technical data defined to include unclassified information useful in 'the design, production, manufacture, repair, overhaul, maintenance or reconstruction of . . . implements of war on the U.S. munitions list' or 'any technology that advances state-of-the-art or establishes a new art in any area of military applicability'. Export in the sense of the regulations includes the disclosure of information orally as well as shipments. There are also controls on publication. Data in published form and subject to public dissemination are excluded from the licensing requirements of the ITAR.

A Defense Department official was recently reported to have said that 'ITAR, if enforced to the letter, would cover virtually everything done in the United States. But people understand they are written very generally'. He later added that the regulations will not be capriciously enforced and pointed out that no prosecutions have yet occurred for the publication of scientific or technical articles.

In December 1980, the Department of State proposed a revision of ITAR that would narrow somewhat the definition of technical data (the part about advancing the state-of-the-art was deleted) and weaken the requirement for obtaining clearance prior to publication. With the change of administration in January 1981, the Department of State was required to consider the proposed changes in light of a redirection of policy. As of December 1982 the department was still reviewing the proposed revisions and expected to publish its updated version in the Federal Register in early 1983 with a request for public comment.

The Export Administration Act of 1979 and Export Administration Regulations

The Export Administration Act of 1979 governs the export of goods and services not covered by the Arms Export Control Act and is the principal statute concerning the export of unclassified technology.

The Secretary of Commerce is authorized to restrict or prohibit certain exports including technologies to protect national security, to implement foreign policy, or to conserve supplies of critical materials or reduce the inflationary impact of foreign demand. The export of goods or technology 'available without restriction from sources outside the United States in significant quantities and comparable in equality to those produced in the United States' are exempt from controls for national-security or foreign-policy reasons unless the President determines that this exemption would be detrimental.

Under the authority of the act the Office of Export Administration of the Department of Commerce administers the Export Administration Regulations (EAR), which include a section on technical data. The data covered are those defined by the act, which states that technology 'means the information and know-how that can be used to design, produce, manufacture, utilize, or reconstruct goods, including computer software and technical data, but not the goods themselves'. Although the export of technology is not defined by the act, the regulations cover not only shipments but also oral exchanges, visual inspection, and the foreign application of the personal knowledge or technical experience acquired in the United States. In principle, therefore, conferences and articles are to be considered as methods for exporting technology.

A Commodity Control List (CCL), which indicates the goods or technology subject to export controls, is established and maintained by the Secretary of Commerce under a provision of the Act. Input for those items in the national-security portion of the list is provided by the Secretary of Defense, who has the primary responsibility for developing a Militarily Critical Technologies List (MCTL), for which 'primary emphasis shall be given to (A) arrays of design and manufacturing know-how, (B) keystone manufacturing, inspection, and test equipment, and (C) goods accompanied by sophisticated operation, application, or maintenance know-how'. Subject to the concurrence of the Secretary of Commerce, the MCTL items are ultimately to be included in the CCL. An 'indexing' provision gives the Secretary of Commerce authority to remove items from the national-security section of the list as performance levels advance, and the items become obsolete.

Coordinating committee for multilateral export controls

The Coordinating Committee for Multilateral Export Controls (COCOM) is the result of an informal agreement among the United

States, its NATO partners (except Iceland), and Japan to coordinate export controls to Communist nations. Established in 1949, COCOM provides participating nations with a forum for the consideration of export controls to Warsaw Pact countries and the People's Republic of China and other Communist Asian nations. It has three major functions listed below.

1. Establish and update lists of controlled products and technologies. Although COCOM lists are not published, they become the basis for the national control lists administered by each member government. A major review of these control lists is underway.
2. Act as the clearing house for requests submitted by the member governments to ship specific items to specified end-users in the proscribed countries.
3. Serve as a means of coordinating the administration and enforcement activities of the member governments.

Officially the three lists of COCOM controlled commodities are classified. However, it is well known and officially acknowledged that the items on the COCOM list are on the U.S. Commodity Control List. With this and other information, knowledgeable sources have partially reconstructed the COCOM lists. The three lists are the following:

1. A munitions list
2. An atomic-energy list including all sources of fissionable materials, nuclear reactors and reactor components
3. An industrial/commercial list, which includes dual-use items with both civilian and military applications.

Because it is fairly clear what items belong on the first two lists and because of the obvious security implications of exporting such items, they cause little disagreement within COCOM. It is the industrial/commercial list that causes most of the controversy and accounts for most of its work.

'The best defense is a flourishing civilian economy'

We can conclude that the broad definitions and the dual-use dilemma of the technologies and exports that fall under the above mentioned control, give the United States government a 'free-ride' in prohibiting almost all kinds of science and technology transfer, which is, obviously one of the first conditions for commercial application. According to D. Bruce Merrifield, the Commerce Department's Assistant Secretary for Productivity, Technology and Innovation, even SDI work that is unclassified, released, and appropriate for commercial applications will not readily yield business benefits to firms that are not SDI contractors, because such research is often 'bottled up by bureaucratic procedures in Washington'. Even the major SDI contractors — including Boeing, TRW, Hughes Aircraft, Lockheed, Rockwell, Teledyne, Brown Engineering and LTV Aerospace — could find themselves impeded by red tape. Bobby Inman, chairman of the Microelectronics and Computer Technology Corp., contends that the Defense Department's long procurement cycle has become a major hindrance to commercialization. Defense research produced numerous spin-offs in the 1950s when the procurement cycle was four to five years, explains Inman, but with the current twelve- to thirteen-year cycle 'the United States loses most of the economic commercial advantage that we once obtained from the early availability of technology'. Increased technical complexity has slowed the process, and so has public and congressional pressure for closer scrutiny of military projects. Although Inman would like to see the bureaucracy streamlined, the current fuss over Defense Department procurement abuses makes that politically impractical. Robert N. Noyce, a coinventor of the integrated circuit and vice chairman of Intel Corp., Santa Clara, Calif., noted that the DoD often takes over promising fundamental technologies for its exclusive use. The Defense Department's VHSIC (very high speed integrated circuit) program, says Noyce 'was originally justified on the basis that it would have commercial potential and benefit. But right now the Defense Department is trying to classify the whole thing. The military tends to go its own path instead of reinforcing the commercial activities'.

Noyce noted that the DoD is currently pursuing a different communications protocol than that being developed by the IEEE and the International Standards Organization: 'It's a completely parallel development which has no spin-off either way', Noyce said. 'What the commercial world is doing can't help the DoD and what the DoD is doing can't help the commercial world'.

The SDI's commercial potential could also be nullified by Government rules barring or severely restricting exports of high technology. These rules have been expanded recently to include not only weapons but also fundamental technologies with no proven military application.

One example of this is biocomputing, in which organic cells supplant sillicon chips as circuitry, adding vast processing capability in a smaller package. Ionson has singled out this still largely theoretical technology as having the potential to create 'an absolute revolution in the computing industry'. But the Defense Department announced recently that it is seeking to apply export controls to the nascent field. Since publication or discussion with a non-United States citizen is considered a form of export, the rules would hamper the dissemination of results from early research as well as the marketing of any products that might eventually emerge.

If the Government applies these controls to all the promising technologies embodied in the SDI, said Representative George Brown (D-Calif.), a member of the House Committee on Science and Technology, 'it will cause trade problems as well as problems improving innovation and productivity in our own country'.

In conclusion, we have raised ample evidence to answer the question of the character of the SDI program. SDI is a research program geared to provide the knowledge required to support a decision on whether to develop and later deploy advanced strategic defensive systems. Except for its small 'mission oriented'-basic research program, SDI R &D is mainly generic by nature, aimed at advancing the basic technology required to ultimately develop specific developed systems. Although this might generate probable civil applications, the exotic nature of this military program will most certainly, and contrary to military and NASA programs in the past, lead to converging need trajectories with civil applications. In addition, the military context of the SDI program and the extensive restrictions on technology transfer, demanding strong restrictions on publication and participation of SDI-scientists in open scientific meetings, reduces the possibility of commercial spin-off.

We are closing this chapter with an appropriate citation from Lester Thurow in an article in *Technology Review*:

In the end one should ask whether those who debate Star Wars are even raising the right questions. None of history's great superpowers — from Egypt through Great Britain — has ever been conquered from the outside by rivals without first declining economically. In many ways, a prosperous American civilian economy is our best military defense. The real economic case against Star Wars is that the civilian economy — battered by trade deficits that have no parallel in human history — could well use some of the time, attention, and money now being lavished on Star Wars.

References

Because of SDI — related work, research can only discuss recent laser advances in general terms, *Research & Development*, April 1986.

Congressional Budget Office, *Federal Support for R&D and Innovation*, April 1984.

Ember, L., 'Star Wars Funds Spurs University Research', *Chemical & Engineering News* 17 February 1986.

Galbraith, J.K., *The New Industrial State*, Boston Houghton Mifflen, 1978.

Ministerie van Defensie, *Nota Defensie Technologie*, Den Haag, February 10, 1986.

News 1986, 'R &D-budget: prosperity for defense, pain for national labs,' *Physics Today*, April 1985.

OECD, *Science and Technology Policy Outlook*, Paris, 1985.

Office of Technology Assessment, *Microelectronics Research and Development*, Background Paper, March 1986.

Reich, R.B., 'High Technology, Defense and International Trade', in J. Tirman (ed.) *The Militarization of High Technology*, Cambridge, Ballinger Publishing Company, 1984.

N. Rosenberg, *A Historical overview of the evolution of federal investment in R&D since World War II*, Paper for Workshop on: The Federal Role in R &D, November 21–2, 1985.

Thurow, L., 'The Economic Case Against Star Wars', *Technology Review*, Febr./March 1986.

Tirman, J., 'The Defense-Economy Debate', in J. Tirman (ed.) *The Militarization of High Technology*, Cambridge, Ballinger Publishing Company, 1984.

Walton, P., 'Star Wars Office kept Fusion Research Secret', *New Scientist*, July 3, 1986.

Part III: Economics of Military R&D

6 SDI and innovation processes

The innovation process

This chapter deals with the process of technological innovation itself. This process can be defined as follows: 'Industrial innovation includes the technical, design, manufacturing, management and commercial activities involved in the marketing of a new (or improved) product or the first commercial use of a new (or improved) process or equipment' (Freeman, 1974). In other words, industrial innovation involves the commercialization of technological change. As our definition suggests, however, innovation does not necessarily imply only the commercialization of a major advance in the technological state-of-the-art (a radical innovation) but it includes also the utilization of even small scale changes in technological know-how.

In practice, of course, radical and incremental innovations often will be closely linked, and a radical innovation will pave the way for an extended series of improvements, the sum total of which can have as marked an influence on the innovation's commercial performance as did the original breakthrough. It might also pave the way for a wave of innovations in other fields as, for example, did the transistor. In competitive terms, while the introduction of a major new combination can afford the originator significant technical and market leads, continued success depends on the ability of the firm continuously to improve the performance of the new device. Technological innovation, generally is a dynamic, iterative learning process rather than a 'one-off' event. This suggests that a successful new design involves only a temporary balance of price and non-price characteristics and continued success requires frequent adjustments to this balance. Moreover, it suggests also that innovation is a process involving not only the innovating company but its customers also, especially during its latter phases.

Some over-simplified early 'models' of innovation, emphasized the causal role of scientific and technological advance and were generally linear; this is the so-called 'technology-push' model of innovation. According to this model, discoveries in basic science lead eventually to industrial technological developments which result in a flow of new products and processes to the market place.

From the early to mid-1960s onwards, largely as the result of a growing number of empirical studies and descriptions of actual innovations, the role of *felt need* in innovation began to be emphasized which led increasingly to the adoption of the linear 'need-pull' model of innovation. Here, innovations arise as the result of a perceived, and often clearly articulated market need.

During the past decade, both the 'pure' technology-push and need-pull models of innovation have increasingly been regarded as extreme, and untypical examples of a more general process of coupling between science, technology and the market place (Mowery and Rosenberg, 1978). In the first case, it is obvious that 'more R&D' has not necessarily resulted in 'more innovation'. In the second case, over-emphasis on market needs can result in a regime of technological incrementalism and lack of radical innovation (Hayes and Abernathy, 1980). Moreover, the relative importance of technology-push and need-pull might vary considerably during different phases of the industry cycle. A more representative model of industrial innovation is the so-called 'interactive model'. According to this model innovation is regarded as a logically sequential, though not necessarily continuous process, that can be sub-divided into a series of functionally separate but interacting and interdependent stages. The overall pattern of the innovation process can be thought of as a complex net of communication paths, both intra-organizational and extra-organizational, linking together the various in-house functions and linking the firm to the broader scientific and technological community and to the market place. In other words the process of innovation represents the confluence of technological capabilities and market needs within the framework of the innovating firm.

Despite the increasing acceptance of the interactive model of innovation, it nevertheless remains clear that many governments — and indeed many industrial companies — continue to adhere, at least implicitly, to the technology-push model, as seems also the case in the SDI program. This is reflected in the emphasis in most countries on funding R&D — often to the neglect of other areas, such as demand — in the belief that 'more R&D' does equate to 'more innovation'. Although it can't be objected that in the SDI case there will be a close connection between 'development' and 'innovation', being a major development program leading specifically to an operational space shield. This leaves, however, the question of the extent to which the results of basic research, in the form of scientific advance, contribute to technological innovations in industry. Is basic scientific understanding a pre-requisite to technological progress? If so, is the gap between scientific advance and industrial technological application decreasing? These questions, which are clearly important from the point of view of public policy, are addressed in the following sections.

The discovery-innovation time lag

An interesting and long-standing debate in the innovation field centres on the question of whether or not the elapsed time between scientific discovery or invention and industrial innovation (commercialization) is decreasing. Mensch (1979), for example, has suggested, on the basis of his observed bunching of innovations during the depression periods of the Kondratiev cycles, that a 'depression accelerator' effect can occur, which shortens the time between invention (including basic discovery) and commercialization. A second, and more general, factor is the rapid growth during the twentieth century of industrial R&D laboratories (Freeman, 1974). This internalization of scientific and technological activity by mainly large companies (who can also forge good contacts with research in universities and government research establishments) greatly increases their ability to capitalize on scientific advance and invention, which might be considered to have great potential for reducing the gap between discovery and innovation.

The question is, of course, has the expected shortening in discovery–innovation gaps occurred? Unfortunately, as we shall see below, the evidence on this point is far from conclusive. Enos, for example, on the basis of his study of 35 innovations, found the following time gaps (data taken from Giarini and Loubergé, 1978):

— Pre 1900 (12 innovations); 12.5 years

— 1911–1940 (11 innovations); 15.8 years

— post 1941 (12 innovations); 12.5 years

Hafstad (1966) has presented data which suggest that a definite shortening has occurred in the discovery/innovation time gap: photography — 112 years; telephone — 56 years; radio — 35 years; radar — 15 years; television — 12 years; atomic bomb — 6 years; transistor — 5 years.

From the point of view of public policy, perhaps the most notable contribution to this debate was that by Cook and Morrison who pointed to the role of need factors in shortening the discovery/innovation gap: 'Even a casual inspection of the examples can leave little doubt that a major factor in shortening the time-gap is the intensity with which the need for the innovation is actually felt by the potential user'.

Cook and Morrison point to four degrees of felt need and provide examples of each:

— Need felt in advance: examples where an urgent and explicit need was felt by the potential user before the innovation and acted to shorten the time gap are the cotton gin, radar, paper and high-yield penicillin.

— Need felt only after the discovery of invention: examples of where the customer did not realize he had a need until after the concept, device, process or system was made, but realized his need quickly and shortened the time gap, are X-rays, wireless, photography, telegraph and hybrid corn.

— Need not felt until long after the discovery or invention: examples in which a lack of explicit-felt need on the part of potential users caused the innovation to progress slowly — usually combined with difficulty in determining or recognizing the technical adequacy and economic attraction of the innovation, or because satisfactory alternatives were already in successful use, thus making the situation strictly a competitive one — e.g. the rocket motor, magnetic recording, catalytic cracking and the fluorescent lamp.

— Negative need, or actual objection: an example in which one group of potential customers felt a negative need and delayed the innovation resulting in a long time gap as the sewing machine.

It might be concluded from the above that public bodies can accelerate the application of scientific and technological advance in certain desired directions both by the clear articulation of unmet public needs — as in the case in the SDI program — and through disseminating information concerning identified needs elsewhere in society. This conclusion is consistent with the results of an analysis by Nelson (1982) who, on the basis of descriptions of the role of the United States government in the development of seven major industrial sectors, emphasized the important role played by public procurement agencies. Finally governments might have a role to play in helping to overcome various societal and institutional barriers to the acceptance of new technology, thereby accelerating rates of commercial application.

The role of scientific advance in technological innovation

While it is clear from the previous section that is is not possible to provide a definite answer to the question of shortening the discovery/innovation time gap, it is clear that scientific advance often does play an important role in technological innovation. However, to what degree science plays a direct role in innovation, or to what degree it plays a more diffuse enabling role, is far from clear. Nor is it by any means certain that scientific advance necessarily precedes technological development and commercial application. Rosenberg suggests that the flow is often in the opposite direction and, if this is true, it renders much of the scientific discovery/application time-lag debate meaningless.

The historian of science, Derek de Solla Price (1965), was one of those who seriously questioned the proposition that scientific advance is a necessary pre-requisite to technological innovation. He argued that science and technology essentially are two separate worlds, that the contact between them is spasmodic and occasional and that technology 'feeds on itself' for long periods without any major influence from science. Indeed, Price pointed out that the 'actors' associated with these two activities (science and technology) are different, as are their aims. Scientists are concerned primarily with generating scientific knowledge. Engineers are concerned with generating artifacts offering commercial benefits. Price used the simile of two dancing partners to illustrate the nature of the science/technology interaction, who both have their own steps even while dancing to the same tune; occasionally they will come together and be in-step, but usually they will be out of step.

Freeman (1974) later took up Price's model and argued that while it might well have been a true description of the science/technology interaction during the nineteenth century, it fails adequately to describe the relationship today. Today's relationship he sees as being much more 'intimate', arguing that the professional R &D department is both a cause and a consequence of this intimacy.

Certainly, given the growing employment of scientists in industrial R &D with their ability to 'plug-in' to an 'invisible college of peers' working in universities and government research laboratories, while at the same time pursuing their own, more directed research interests, Freeman's claim would appear to have some validity. The question remains, of course, as to whether or not we can muster sufficient empirical evidence to support or refute this claim.

During the 1960s two well-known and detailed attempts were made in the United States to determine the role that fundamental, 'non-mission oriented' research plays in technological innovation. These are TRACES (Illinois Institute of Technology, 1968) and Project Hindsight (Sherwin and Isensen, 1966). The former investigation was based on tracing key scientific events which contributed towards five major technological innovations. The latter was a study of 'recent' science and technology that had been utilized by the United States Department of Defense in the development of twenty post Second World War weapons systems.

The results of TRACES and Hindsight are contradictory. According to TRACES, 70 per cent of the 'key events' which contributed to the five major innovations studied were 'non-mission research events'; according to Hindsight 'undirected' (non-mission or fundamental) research played an extremely minor role in the development of the twenty weapons systems. However, the apparently minor role played by undirected research reported in the Hindsight study is almost certainly due largely to the fact that the investigation began with 1940 and deliberately chose to exclude the 'pool of basic knowledge' assembled before 1940. The significance of this omission is demonstrated by the results of TRACES which showed that 90 per cent of all non-mission research events were complete at least ten years before the innovation. Clearly, the starting point is crucial.

It might be pointed out, in addition, that since TRACES was funded by the National Science Foundation, an organization responsible for funding 'non-mission research' in the United States; it might not be altogether surprising that such research was found to be crucial to technological innovation. Hindsight, on the other hand, was funded by the Department of Defense, an organization involved in funding 'mission-oriented' research and technological development; again, perhaps not surprisingly, the results confirmed the importance of such 'directed' R&D. Moreover, TRACES was almost certainly undertaken in response to the results of the Hindsight project.

Another US study, this time undertaken by the United States National Academy of Science (1965), yielded results supporting those of TRACES. The NAS study investigated scientific publications which announced 'practical discoveries' in chemistry, and the basic research results which led to these 'discoveries' were, on the basis of cited references, traced back to their origins. Publications relating to sixteen different industrial discoveries included two-hundred and forty citations; 65 per cent of these referred to university research, 31 per cent to industrial research and 4 per cent to other sources. Further, 67 per cent of the citations referred to fundamental science journals and books, 22 per cent to applied journals, 10 per cent to patent publications and 1 per cent to other sources.

Few, of course, will be surprised to learn that basic scientific research makes a significant contribution to technological developments in chemicals, a science-based industry which itself performs basic research. A study in the UK, however, yielded results which suggested that university chemistry research has played only a relatively minor role in industrial chemical innovation. The study set out to determine the relationship, over time, between university research in organic chemistry and the use by British industry of the results of this research (Langrish, 1974). The technique used was to study the institutional origins of abstracts produced by the Journal of the Society of Chemical Industry, which for 80 years or more has produced abstracts of the world's literature that might be of relevance to industrial chemists.

The study showed that there had been a marked change over time in the main institutional sources of abstracts, from European universities (mainly German) in 1884 to American industry in 1952. Since the end of the nineteenth century foreign industry has been the major source of inputs with, in 1952, United Kingdom industry being the second most important source. The United Kingdom university has consistently been the least, or the second least, important source of inputs.

Langrish offered two alternative hypotheses as to why the importance of university research has apparently declined:

(1) Industry has increasingly taken over its own research;
(2) A new branch of science is only useful to industry in its early days. (For example, the early days of astronomy as a science were linked with economically important attempts at improving navigation but it has hardly been useful since then.)

Langrish then states: 'The relationship between university research and industry may well be a function of the degree of development of the area concerned. Once a new area has been established, the aim of science is to understand; the aim of technology is to make it work, and industry has been very successful at making things work without too much reliance on understanding'.

On the basis of Langrish's data for the chemical industry, it might be postulated that in certain areas of science the role of universities is to make the initial fundamental breakthrough, followed by many years of basic research to understand the nature of the process; the role of industry is commercially to utilize the breakthrough, concentrating on understanding and harnessing its effects while being largely unaware of their causes. Given the very rapid growth in industry R&D, moreover, it might also be suggested that without a very considerably in-house R&D capability, industry would be largely unable either to utilise the potential of the breakthrough or progressively to enhance this potential. Industry has thus needed increasingly to take over the relevant fields of research itself. At the same time the universities have gone on to open up new fields of research, perhaps paving the way for the next generation of industrial applications.

The results of a study by Gibbons and Johnston (1974), in contrast to those of Langrish, suggested that university scientific research in the United Kingdom does make a significant direct contribution to industrial R&D; it is, perhaps, this direct input that is not captured by Langrish's citation technique. Gibbons and Johnston identified (in the early 1970s) 887 units of information input that contributed to the solution of technical problems encountered during 30 recently completed or ongoing industrial innovation projects. 300 of these information inputs were from external sources, of which 107, or 36 per cent, could be classified as 'scientific' information. Of these 107 scientific inputs, 36 came from the scientific literature (which might be expected to contain a strong university representation, i.e. an indirect university input), 8 from scientific textbooks, 11 from scientific handbooks, 4 from scientists met at conferences, 18 from scientists in government research establishments and 30 from scientists in universities. Thus, about 10 per cent of the total external technical problem solving information inputs to the 30 industrial R&D projects derived directly from university scientists.

Clearly, the important issue of the role of university research in industrial technological change requires further careful research, preferably on a research field by research field basis, or on a sector by sector basis. Emerging technologies as employed in SDI might be an interesting sample here.

Turning back to the United States, in a follow-up study to TRACES, 'The Interactions of Science and Technology' (NSF, 1973), a study similar to TRACES, was undertaken but which went very much further and analysed the factors which are conducive to successful technological innovation. Ten major innovations were studied in detail, and the relative numbers of 'non-mission oriented research' events, 'mission orientated research' events, 'development research' events and 'non-technical' events which contributed to the innovation. The results indicated that of more than five hundred 'significant' events (i.e. 'an occurrence judged to be important to the development of the technical innovation or its further improvements, as reported in a published paper, presentation or reference to research'), about 40 per cent were non-mission orientated, about two-fifths were mission orientated, about one-fifth were development. Of the ninety 'decisive' events ('occurrences that provide a major impetus to the innovation or its further improvement . . .') about one-sixth were non-mission orientated, about half were mission orientated, about two-fifths were development events and 1 per cent were non-technical events.

We can see from the results of 'TRACES II' that the study attempted to take into account the relative significance of the different 'events', i.e. 'significant' and 'decisive'. While non-mission research made an important contribution to 'significant' events, mission-oriented (applied R&D) made the major contribution to the 'decisive' events.

The question has to be asked whether it is meaningful, with the complex interactive process of technological innovation, to talk of identifying discrete 'events' and attempting to measure with any degree of precision their relative significance. Perhaps the most telling comment on this point was made by Mowery and Rosenberg (1978);

> Indeed, the pervasiveness of uncertainty in the innovation process is ignored by most of the empirical studies. This is particularly true of the studies that have as their unit of analysis some group of selected innovations, and that attempt to reconstruct the history of such innovations: Hindsight, TRACES and the Battelle study are prominent examples. The attempt to decompose neatly the complex, stochastic, and uncertain process which is that leading to innovation into a set of events that can simply be cumulated to yield an innovation, is as gross an over-simplification as is the 'black box' approach of the surveys of business firms. To attempt such a reconstruction, and further to attempt to ascribe relative importance to the various categories of research events, seems fallacious.

Without doubt, the seminal contribution to the debate on the relationship between science and technology, is the paper 'How exogenous is science?' by Nathan Rosenberg which is included in his book *Inside the Black Box* (Rosenberg, 1982). The core theme of this paper is 'that technological concerns shape the scientific enterprise in various ways'.

Of course, the underlying assumption in the various studies described is that the nature of progress is such that science tends to influence technology, rather than vice versa, because scientific understanding precedes technological advance. Rosenberg questions this widely held assumption:

> One of the more misleading consequences of thinking about technology as the mere application of prior scientific knowledge is that this perspective obscured a very elemental point: technology is itself a body of knowledge

about certain classes of events and activities: it is not merely the application of knowledge brought from another sphere. It is a knowledge of techniques, methods, and designs that work, and that work in certain ways and with certain consequences, even when one cannot explain exactly why. It is therefore, if one prefers to put it that way, not a fundamental kind of knowledge, but rather a form of knowledge that has generated a certain rate of economic progress for thousands of years. Indeed, if the human race had been confined to technologies that were understood in a scientific sense, it would have passed from the scene long ago. (Rosenberg, 1982, p. 143).

According to Rosenberg, there are many areas even today where technological progress occurs in the absence of a full understanding of the underlying scientific principles. In these cases, technological change is largely an ad hoc process in which knowledge is accumulated through trial and error. A good example of this is the aeroplane, which has developed despite the lack of any adequate theories of turbulence and compressibility. In other words technological advance largely determines the scientific agenda.

Given the economic incentives underlying technological advance, this pattern is, according to Rosenberg, hardly surprising. Of course, Rosenberg does not argue that technology always precedes science, but that empirical evidence suggests that in general this has been the case. It might be, of course, that gaining a basic scientific understanding of an established technological phenomenon will greatly enhance the scope and accelerate the pace of technological change. For example, copper-oxide and silicon rectifiers were being used before Shockley's research threw light on the phenomenon of electronic conduction in doped semiconductor materials, out of which grew the modern semiconductor industry; animal breeders had succeeded in producing many different forms of livestock and domestic animals before the discovery of DNA paved the way to basic scientific understanding of the possibilities of genetic engineering.

Push or pull?

It should be clear from the foregoing sections that we subscribe neither to the simple linear 'technology-push' nor the simple linear 'need-pull' models of innovation. Technological innovation is, as Mowery and Rosenberg (1978) have shown, generally too complex a process to be described by a simple chain of causality in one direction or the other. It is, nevertheless, worth briefly outlining the major arguments in this area, less from the point of view of specific products or processes, but rather more from the viewpoint of the evolutionary dynamics of new generic technologies or industrial branches.

It was probably the work of Schmookler (1966) which pointed most strongly to the importance of demand factors in innovation. Based on a highly detailed study of the evolution over time (1840–1950) of patents and investment mainly in four capital goods industries in the United States (railroads, petroleum refining, agricultural machinery and paper making) — but also, although to a lesser extent, in a number of consumer goods industries — Schmookler's work led him to the conclusion that market growth and market potential were the main determinants of the rate and direction of inventive activity.

The work that stressed most strongly the role of technology in innovation and economic change, was that of Schumpeter (1943). In the first case Schumpeter stressed the importance of exogenous science and invention which, via the medium of entrepreneurial activity, led to the growth of new industrial branches and new areas of demand (the Schumpeter I model); in the second case he emphasized the role of endogenous science and technology in the R&D laboratories of major companies — strongly coupled to exogenous science and technology — which again led to new patterns of production and new market structures (the Schumpeter II model).

We are thus faced with a dilemma. Two eminent economists, both operating with detailed time series data, reach apparently diametrically opposed conclusions. The question is, how to resolve this dilemma? Fortunately a detailed piece of research, undertaken in the late 1970s at the Science Policy Research Unit of the University of Sussex provided some of the answers. The research, carried out by Walsh, Townsend, Achilladelis and Freeman (1979), looked at an industry that was more science and R&D intensive than those considered by Schmookler, namely the chemical industry: more specifically they looked in detail at four subsectors; plastics, dye-stuffs, pharmaceuticals and petrochemical intermediates. The indicators employed were scientific papers, patents, investment and production.

Walsh *et al* set out to test four 'models' of innovation: the demand-led model (after Schmookler); the model of technology and demand-led science (after Hessen, 1931); Schumpeter's model of entrepreneurial innovation (Schumpeter I); and Schumpeter's model of large firm managed innovation (Schumpeter II).

The most significant result of Walsh *et al's* study, is the evidence of 'two-way linkages with strong reciprocal influences'. In some cases demand might lead to technological activity (and production), followed by basic research to understand the fundamental mechanisms involved (a sort of 'Rosenberg model'); in other cases the results of prior scientific research might be a necessary prerequisite for inventive and then innovative activity, which might then be followed by further basic research better to understand the mechanisms involved. Walsh *et al* terms this derivative scientific activity 'gap-filling' science.

In general, Walsh *et al* see their overall findings as being more consistent with the theories of Schumpeter than with those of either Schmookler or Hessen.

Thus we might conclude from the work of Walsh *et al* that the relationship between science, technology and the marketplace is rarely unequivocally undirectional, nor is it a simple one, and within particular branches of industry causality can switch from being mainly in one direction to being mainly in the other. The linkages between science, technology and the marketplace are complex, interactive and multidirectional, the dominant driving force varying over time and between one branch of industry and the next. Innovation is, as we suggested earlier, a process of coupling, and it should be a primary aim of policy to forge the necessary links, adjusting the balance of resources to match the pattern of requirements for each technology or industry at its particular stage of development.

Technological trajectories

Most of the emphasis in the literature on technological innovation during the 1960s and 1970s was on innovations at the level of the firm. This reflected a growing interest on the part of researchers and policymakers alike in understanding the managerial and other factors associated with successful and unsuccessful innovations.

By and large most of those empirical studies were descriptive and lacked theoretical underpinning.

Economists, in contrast, continued to take a rather macro view of technological change, dealing mainly with process changes and attempting to measure the contribution of 'technical change' to manufacturing productivity growth. In this case technological change, and indeed the firm, were treated as a 'black box' and behavioural and organizational factors largely were ignored.

During the 1970s Richard Nelson and his colleagues at Yale University began the search for a theory of innovation that might link together better the micro and macro worlds that previously had been treated largely as separate entities. This work culminated in the paper 'Towards a Useful Theory of Innovation' published by Nelson and Winter in 1977. In their paper, Nelson and Winter developed the concepts of 'natural trajectories' and 'selection environments'. Taking first natural trajectories, they stated that: . . . it may be that there are certain powerful intra project heuristics that apply when a technology is advancing in a certain direction, and payoffs from advancing in that direction that exist under a wide range of demand conditions. We call these directions "natural trajectories". If natural trajectories exist, following these may be good strategy' (Nelson and Winter, 1977).

Thus the rate and direction of technological advance might be determined not only (as economists would contend) by demand conditions but also, as Rosenberg (1972) puts it, through the operation of certain 'technological imperatives'. In the semiconductor industry, for example, we can see that both technological and market forces drove semiconductor developments along the paths of more reliable, smaller, less energy intensive and cheaper devices. Clearly these developments satisfied the evolving needs of the aerospace, military and computer markets well, but it would have been surprising if accumulated technological and production learning had not moved developments along similar paths.

Nelson and Winter make the more general point that: 'While natural trajectories almost invariably have special elements associated with the particular technology in question, in any era there appear to be certain natural trajectories that are common to a wide range of technologies. Two of these have been relatively well identified in the literature: progressive exploitation of latent scale economies, and the increasing mechanization of operations' (Nelson and Winter, 1977).

A good illustrative example of natural trajectories can also be found in the pattern of developments in textile machinery during the post war era. The principal development trends during this period have been geared towards three main ends: increasing the productivity of individual units of production; reducing the number of operations required in a given production sequence; increasing the amount of automatic transfer between operations (Rothwell and Zegveld, 1979). The many radical and incremental innovations in textile machinery during the past thirty or so years have been directed largely towards satisfying these ends (Rothwell, 1976) which, between them, have established the agenda of both public and private R&D programs.

Turning now to selection environments, Nelson and Winter stated: 'We propose the concept of a "selection environment" as a useful theoretical organizer. Given the flow of new innovations, the selection environment (as we are employing in the terms) determines how relative use of different technologies changes over time. The selection environment influences the path of productivity growth generated by any given innovation, and also it feeds back the influence strongly of the kinds of R&D that firms and industry will find profitable to undertake' (Nelson and Winter, 1977).

Nelson and Winter point to two basic classes of selection environment; market and non-market. In the first case, not surprisingly, the nature of demand acts as a filter for selecting projects for development from a larger portfolio of possibilities. Firms will innovate in areas which yield them the maximum appropriability of benefit from their R&D efforts. In the second case, the motivations of firms in a non-market sector, according to Nelson and Winter, cannot simply be presumed to be monetary profit, and they offer as examples the relationship between a doctor and his patient.

A second non-market selection element that can influence the choice of technology (as well as its rate of development) is government regulations. These can effectively filter out harmful developments and accelerate the search for more socially desirable alternatives (e.g. the discontinuance of development and marketing of cyclamates in the United States and the shift in development effort towards 'natural', high-intensity non-sucrose sweeteners). There does exist compelling evidence that in some branches of technology (e.g. pharmaceuticals) government regulations can have a powerful influence on both rate and direction of technological change (Rothwell, 1979).

The work of Nelson and Winter does, we believe, indeed lay the foundations of a useful theory of innovation. The concept of natural trajectories of technological change (technological trajectories) aids us in taking not so much a macro, but rather a 'meta' view of technological change, enabling us to look beyond what is happening at the level of individual projects and firms. The concept of selection environments assists us in understanding why, given a set of possible trajectories within a rather broad technological regime, certain pathways are preferentially followed. It is important also to note that public institutions and public policy are themselves potentially important elements of the selection environment.

Technological interrelatedness and intersectoral flows

In his paper Technological Interdependence in the American Economy, Rosenberg states:

> Innovations hardly ever function in isolation. Time and again in the history of American technology, it has happened that the productivity of a given innovation has turned on the question of the availability of complementary technologies. Often these technologies did not initially exist, so that the benefits potentially flowing from invention A had to await the achievement of inventions B, C or D . . . The growing productivity of industrial economies is the complex outcome of large numbers of interlocking, mutually reinforcing technologies, the individual components of which are of very limited economic consequence by themselves. The smallest relevant unit of observation, therefore, is seldom a single innovation but, more typically, an inter-related clustering of innovations. (Rosenberg, 1979).

Rosenberg offers a number of examples of technological interdependence, highlighting instances where the practical use of a particular invention was delayed until a complementary development(s) enabled its implementation:

— it was not until the development of cheap, high-quality steel that the potential of the compound steam engine was realized;
— it was not until the development of suitable machine tooling methods that alloy steels realized their full potential;
— many advances in metallurgy were required before the jet engine could be put into general use.

To the above we might add the crucial 'enabling' contribution that computer-aided design made to the rapid development of large scale integrated circuits (the development of CAD systems themselves owing much to previous developments in computing and in semiconductors) and the need for the development of new and modified forms of textile processing machinery before the commercial potential of synthetic fibres could be realized by textile manufacturers.

Finally, it is worthwhile briefly examining a further technological linkage mechanism between users and producers, and that is the role of the user in the innovation process. It was, without doubt, the work of Eric von Hippel that first focused attention on the role of the user in the innovation process (von Hippel, 1976, 1978, 1979). Von Hippel's researches in the scientific instruments industry showed that, from a total of 111 innovations, 76 per cent originated as the result of inventive activity on the part of the user. Shaw (1983) found a similar pattern for innovations in the UK medical instruments industry: in a sample of 33 innovations, 25 were transferred from the user to the manufacturer via a process of continuous and multiple interaction.

Since much of the state-of-the-art research in many fields and in many countries is performed in universities, government laboratories and other public sector institutions, clearly the above results have implications for public policy. Specifically, they again highlight the need for effective linkage mechanisms to facilitate the transfer of user-inventions to manufacturing companies.

Factors in success and failure

The literature on factors affecting the success and failure of innovations is simply too extensive to be comprehensively covered here and we will offer a brief summary of some of the more important points that have emerged.

Overall, there are no magical formulae for innovatory success, and it is more or less possible to sum up the results of the various studies by saying that success is a matter of 'good common sense' and 'overall management competence'. We shall deal with a number of the success factors separately below.

Understanding and meeting user needs

The need to gain an imaginative understanding of user needs, and of interpreting them in the design of a new product or process, is obvious. Nevertheless it is evident from the many empirical studies in this area that firms very often are deficient in this respect. For whatever reason, they are simply unable to establish the optimum performance/price combination from the point of view of the eventual user, and this is the most frequently quoted cause of failure.

Effective coupling with the market-place

In the first case this is essential in establishing the optimum performance/price combination. In some industries it means actually involving the customer in the development process. Indeed in certain industries, as we have seen, the user will often play a very significant and sometimes the major role in invention and early prototype construction. It is also important, of course, to keep in touch with supplier innovations in materials and components.

Effective internal coupling

This is especially important with respect to linking effectively the R&D and marketing functions; in other words ensuring that development activity is clearly directed towards meeting user needs. It is also important with respect to linking the design and development function to the manufacturing function in order to achieve easy manufacture, (Rothwell, 1980): in other words, ensuring products are designed so that they can be efficiently and reliably manufactured, and it is this one area in which the Japanese, for example, are particularly strong. (Schoenberger, 1983).

Effective coupling with external sources of scientific and expertise

Whilst there is no substitute for in-house technical skills, it is nevertheless important to complement these skills when necessary with external expertise. The ability to identify external expertise in specific areas and to do this rapidly can often make a crucial contribution to innovatory success.

Gifted and committed entrepreneurs

These are the so-called product champions, business innovators and technological gatekeepers (Rothwell *et al.*, 1974; Rothwell, 1975). The product champion enthusiastically supports the innovation, especially during critical phases. He remains committed to it when things appear to be going wrong and others are thinking of abandoning ship. The business innovator plays the key managerial role of integration, coordinating the efforts of those involved in the project especially from functionally separated departments. The technological gatekeeper plays an invaluable role in bringing technological information into the firm and in disseminating it throughout the firm. Whether or not these key individuals can operate effectively is a function of both corporate culture and structure.

Efficient manufacturing procedures

The importance of eliminating technical bugs before commercial launch is obvious. Nevertheless it is a message that did not appear to reach many would-be innovators to their cost, and again it is an area in which the Japanese have great comparative strengths.

Efficient after-sales service and user education

Selling a new product is not simply a matter of transferring a piece of technology, but rather of offering an economic package. Servicing and spares supply are important components of this package; so too is user education, and it is crucial to train users in the right uses and limitations of a new device, otherwise if they do misuse it and it malfunctions they will, with some justification, blame the supplier and not themselves.

The above factors are concerned with what the firm actually did or did not do during the innovation project; in other words they are all project execution variables. To these we can add a second set of factors that outline essentially the pre-conditions for corporate innovation, i.e. that define a corporate framework in which innovation can take place.

The first of these factors is top management commitment to, and visible support for, innovation. Top management visibility is absolutely crucial, especially in the case of major innovations, to overcoming the barriers and resistances to innovations that often exist in companies. For example, people committed to one area of production might resist moves to new areas and it is a top management function to facilitate such shifts. This is especially important within the framework of reindustrialization.

The second factor is the importance of having a long-term corporate strategy in which innovation plays a key role. Innovation should not be an ad hoc process, but one that has direction and purpose. It might be that firms are sometimes compelled to innovate in response to unexpected competitive actions, in response to sudden market shifts, or in response to the emergence of a significant new technological capability, but such cases should be the exception rather than the rule. Firms need strategy in order to obtain synergies between R&D projects, manufacturing projects and marketing projects. Properly planned, one project can contribute to the next project and so on. In other words a coherent strategy enables firms to build on past successes and to capitalize effectively on emerging strengths.

The third factor is the need for an associated long-term commitment to major projects, based not on short-term return on investment criteria, but on considerations of future market penetration and growth. There is a good deal of evidence to suggest that many firms, particularly in the United States and the UK, increasingly have adopted a 'cash flow' view of development activity, and as a consequence have focused increasingly on short term projects that yield quick returns, while innovation, and especially major innovation, requires a longer-term view, and at least some of the firm's projects should be funded with 'patient' money (Hayes and Abernathy, 1980).

The fourth factor is corporate flexibility and responsiveness to change. In some industries production sequences have tended to become increasingly specific with respect to the production of dominant product designs. This means that they become more and more rigid and less able to accommodate product changes. In other words, because of increasing inflexibility in production, it is often seen as being too expensive to introduce product improvements. It seems likely that the introduction of flexible manufacturing systems will largely overcome this problem, enabling innovative firms to offer not just a single dominant design, but a 'family' of designs having a broad range of operation characteristics to appeal to a wide range of customers. We should also recognize that technological innovations might require organizational and market innovations to facilitate their implementation, and again this is especially the case with radical innovations. Management must accept this and the company should be sufficiently flexible to accommodate it.

The fifth factor is top management acceptance of risk. Innovation is inherently a high risk undertaking and one of the few things we can be sure about is that there will be failures. Management must accept this, and not use one failure as an excuse for withdrawing from the innovation race altogether. In addition, attempts should be made to learn the lessons of failure through analysing unsuccessful projects, which many firms fail to do. Associated with this is the need for termination criteria. Potential failures can continue for a considerable period under their own momentum with extremely high opportunity costs for more promising projects, and it is essential that top management accept the responsibility for the termination of failing projects.

The final factor is the creation of an environment within the firm in which entrepreneurship can flourish and innovations occur. The literature on this point is copious but it appears to point to two distinct modes of organization, one conducive to, and one inimical to, in-house entrepreneurship; these have been termed 'organic' and 'mechanistic' respectively.

A significant fact to emerge from these innovation studies is that, outside the United States (where public procurement and regulations have been important), governments have rarely been seen to play a major determining role in either success or failure. In some instances this has undoubtedly been a function of the research methodologies employed, which have often concentrated solely on firm-specific factors. It might also point to a more fundamental factor, which is that innovatory success or failure, in practice, is determined largely by the actions of management. The implication of this, of course, is that in the case of specific innovation projects in firms, there are inherent limitations to what innovation policies can achieve. While governments can, to be sure, create a national environment conducive to innovatory endeavours, while they can supply certain inputs to the innovation process and act as an innovation-receptive market, and while they can remove certain barriers to innovation, in the final analysis success depends on the propensities and abilities of industrial managers. In other words, while governments can act to complement the innovatory endeavours of technically progressive managers (establish the necessary 'enabling conditions'), they can do little in the face of incompetence or simple indifference. Policy makers would do well to bear this in mind.

Finally, all the major studies of success and failure in innovation greatly emphasize the importance of intra-firm and extra-firm linkages. They support the interactive model of innovation as a process of communication. Establishing efficient internal linkages is, of course, a prime function of management, as is establishing appropriate mechanisms for linking the firm to the marketplace and to external sources of scientific and technological expertise and advice. With regard to the latter, moreover, there exists an important role for public policy in facilitating linkages between the publically-funded scientific and technological infrastructure and industry.

In summary it can be stated that although analysis of innovation processes would indicate the importance of the development of emerging technologies within the framework of SDI, this is no guarantee for economical benefits of these technologies. As we have illustrated in this chapter the success of industrial innovation depends only for a small part on the results of generic, basic research. The failure or success of industrial innovation is the result of an interaction process between the technical, design, manufacturing, management and commcercial activities involved in the marketing of a new or improved product or process. Enlarging the innovative performance of industry can be stated to be a much more complex problem than the development of these technologies *per se*.

Moreover, as it was stated in a recent article in the *Wall Street*

Journal by Mr. Wolf, director of Rand Corp's Research Programme in International Economics and Dean of Rand's Graduate School:

The Europeans hope technologies resulting from SDI-related programs will provide a stimulus to disappointing rates of growth in output and employment of their ailing economies. The sober truth is that such hopes are distinctly unwarranted and unrealistic. The ills that beset the Western European economies won't be allayed, let alone cured, by participation in a massive state project. An SDI-type effort shouldn't be sold on the basis of false claims, nor bought on the basis of unrealistic hopes. Western Europe's main economic problems derive from structural and institutional conditions quite independent of Star Wars technologies and most unlikely to be affected by them. If these underlying causes of Western Europe's economic difficulties are remedied, the sclerotic performance of the European economies will be improved. If the causes are not remedied, the outlook will remain clouded, and neither SDI nor 'Eureka' is going to do it. If those programs' supporters encourage or indulge unrealistic hopes, disappointment and recriminations will ensure, thereby further weakening the NATO alliance (Wolf, 1985).

References

Cook, L.G., and Morrison, W.A., *The Origins of Innovation*, Report no. 61-GP-214, General Electric Company, Research Information Section, New York, 1961.

Freeman, C., *The Economics of Industrial Innovation*, Penguin Modern Economics Texts, 1974.

Giarini, O. and Loubergé, H., 'The Diminishing Returns of Technology', *Systems Science and World Order Library*, Pergamon Press, 1978.

Gibbons, M. and Johnston, R., 'The Roles of Science in Technological Innovation', *Research Policy*, Vol. 13. November 1974.

Hafstad, L., 'The Role of Industrial Research', *Science Journal*, September 1966.

Hayes, R.H., and Abernathy, W.J., 'Managing our way to Economic Decline', *Harvard Business Review*, July-August 1980.

Hessen, B., 'The Social and Economic Roots of Newton's Principle', in N. Bukharing, (ed.), *Science at the Crossroads: Papers from the Second International Congress of the History of Science and Technology*, (revised edition), London, F. Cass, 1931.

Von Hippel, E., 'The Dominant Role of Users in the Scientific Instrument Innovation Process', *Research Policy*, Vol. 45, No. 3, 1976.

Von Hippel, E., 'Users as Innovators', *Technology Review*, Vol. 80, No. 3, January 1978.

Von Hippel, E., *Appropriability of Innovative Benefit as a Predictor of the Functional Locus of Innovation*, Sloan School of Management, MIT, Working Paper 1084–79, June 1979.

Illinois Institute of Technology, *Technology in Retrospect and Critical Events in Science (Project TRACES)*, Washington, DC, NSF-C535, 1968.

Langrish, J., 'The Changing Relationship Between Science and Technology', *Nature*, 23 August 1974.

Mensch, G., *Stalemate in Technology*, New York, Ballinger, 1979.

Mowery, D.C. and Rosenberg, N., 'The influence of Market Demand Upon Innovation: A Critical Review of Some Recent Empirical Studies, *Research Policy*, Vol. 8, April 1978.

National Academy of Science, *Chemistry: Opportunities and Needs*, NAS, Washington, 1965.

National Science Foundation, *The Interactions of Science and Technology*, NSF, Washington, 1973.

Nelson, R., (ed.), *Government and Technical Change*, New York, Pergamon Press, 1982.

Nelson, R., and Winter, S., 'Towards a Useful Theory of Innovation', *Research Policy*, Vol. 6, 1977.

Nelson, R. and Winter, S., *An Evolutionary Theory of Economic Change*, Cambridge, Mass./London, 1982.

Price, D., 'Is Technology Historically Independent of Science?', *Technology and Culture*, Vol. 6, No. 4, 1965.

Rosenberg, N., *Technology and American Economic Growth*, Harper Torch Books, New York, 1972.

Rosenberg, N., 'Technological Interdependence in the American Economy', *Technology and Culture*, January 1979.

Rosenberg, N., *Inside the Black Box*, Cambridge University Press, 1982.

Rothwell, R., 'Intracorporate Entrepreneurs', *Management Decision*, Vol. 13, No. 3, 1975.

Rothwell, R., 'The Sulzer Weaving Machine: A Case Study of Successful Innovation', *Textile Institute and Industry*, May 1976.

Rothwell, R., *Government Regulations and Industrial Innovation*, Report to the Six Countries Programme on Innovation, P.O. Box 215, 2600 AE Delft, The Netherlands, 1979.

Rothwell, R., 'It's not (just) what you make, it's how you make it', *Design*, (Design Council, London), March 1980.

Rothwell, R., Freeman, C., Jervis, P., Horsely, A., Robertson, A.R. and Townsend, J., 'SAPPHO-Updated: Project SAPPHO Phase II', *Research Policy*, Vol. 13, No. 3, 1974.

Rothwell, R., and Zegveld, W., *Technical Change and Employment*, London, Frances Pinter, 1979.

Schmookler, J., *Invention and Economic Growth*, Cambridge, Mass., Harvard University Press, 1966.

Schoenberger, R.J., *Japanese Manufacturers Techniques: Nine Hidden Lessons in Simplicity*, The Free Press, 1966.

Schumpeter, J.J., *Capitalism, Socialism and Democracy*, New York, Harper and Row, 1943.

Shaw, R., *The Role of the User in the Generation of Innovations in the UK Medical Equipment Industry*, (Mimeo), School of Management, Polytechnic of Central London, Marylebone Road, March 1983.

Sherwin, W.S. and Isensen, R.S., *First Interim Report on Project Hindsight*, Washington, CFSTI, 1966.

Walsh, V., Townsend, J., Achilladelis, and Freeman, C., *Trends in Invention and Innovation in the Chemical Industry*, Report to SERC (Mimeo), Science Policy Research Unit, 1979.

Wolf, Jr., C., SDI is no European Economic Elixer, *The Wall Street Journal*, 23 October, 1985.

7 Economic benefits of SDI: spin-offs and crowding out

An important issue in the debate on the pros and cons of the SDI program is the likely level of spin-off. What will be the eventual civil benefits of military spending and what is the relationship between government R&D spending in the military sector and economic growth?

The frequency with which these questions on the legitimation of military R&D spending are posed can be directly related to the size of the defense R&D budget. Historically we find a peak in the spin-off debate in the United States in the late 1950s and the early 1960s at the time of the Apollo project, when military spending was nearly 80 per cent of the federal R&D-budget. At the same time, many studies on the relationship between military R&D and economic growth were undertaken. The spin-off debate diminished in the 1970s and re-emerged in the 1980s when selective growth and civil goals influenced patterns of R&D spending. The rise of the military R&D-budget, and especially SDI, has again revived the spin-off debate.

A simular situation can be found in the Federal Republic of Germany (FRG) where the 'spin-off argument' played an important role in the reconstruction of the 'Bundeswehr' in the period from the early 1950s till the mid-1960s. The broad diffusion of this argument shows, in this period, that the stimulation of economic growth by means of a strong, R&D-intensive defense-industry was a central part of the science and technology policy of the FRG. This strategy was replaced in 1966/7 by a concept which focused on direct governmental policy influence on industrial R&D.

In the 1970s the spin-off argument was hardly used in West Germany and in governmental publications on R&D and defense as a policy instrument it largely disappeared. The 'Bundesverband den Deutschen Industrie', which, in the 1950s and 1960s stressed the useful side-effects of military R&D, put this earlier concept aside in the late 1970s. It was then that the Government of the FRG asked industry to financially support new R&D investment, while pointing to possible spin-offs. At the same time it can be observed that, since 1980, the argument was once again taken up in science-policy circles, especially by policy decision-makers. The same change occurred in the leading defense journals and the defense industry. Only in the debate on SDI has the federal government of the FRG, for the first time for ten years, again adopted the spin-off argument to support the defense and national security policy with industry and science policy arguments (Rilling, 1985).

Pros and cons in the spin-off debate

The benefits to the civil sector from large military R&D efforts have long been a matter of much debate. Countries which do not have such military R&D efforts tend to view the latter as a form of government support to industry which gives significant advantages in high technology areas. Conversely, those which have such programs tend to regard them as a necessary defense burden which may contribute some spin-off to the civil sector but may also diminish a country's competitiveness by diverting resources away from commercial objectives.

In the SDI-debate, as presumably, at the time of the Apollo project, many managers and supporters of SDI foresee significant commercial benefits arising in the near future. Their arguments are that the volume of SDI is so great that civil benefits are bound to occur. SDI's Organization Director, James Abrahamson, stated that the volume and technological challenges of SDI greatly exceed those of the Apollo program. The project is totally directed to 'technologies of the future': information technology, laser technology and space technology. These advanced, fundamental technologies, in particular should have significant spill-over to civil developments. A third argument is that there exists a general trend of an increasing overlap between civil and military high technology, which means that the expected spin-off will be inherently extensive. Nuclear physicist Edward Teller, for example, estimates a 90 per cent commercialization rate for SDI projects, while President Reagan himself estimates that the technical and economical benefits of SDI will exceed the costs by a factor of fourteen (van Bentum, 1986).

SDI officials are quick to assert that the value of the program does not depend on its civilian spin-offs, but they are also eager to use the lure of these benefits to win support for the ambitious undertaking. Indeed, Abrahamson created an Office of Educational and Civil Applications last fall to promote commercial and other uses of SDI research.

A SDI missile defense system will require major advances in many different areas of military technology, e.g. sensors, high speed missiles, lasers, particle beam accelerators, control devices, power sources, materials and computer hardware and software. Converting research in such areas into saleable products will not be easy. Companies will have to run a gauntlet of classification procedures, export controls and slow development cycles before they can hope to cash in on SDI research results. Perhaps not surprisingly, Defense Department insiders are generally optimistic. For example, SDI chief scientist, Gerold Yonas, points to space surveillance research, which accounts for 35–40 per cent of SDI spending, as an area in which 'spin-offs are expected into every part of the economy'. Brigadier General Robert Rankine, Air Force special assistant for SDI, additionally believes that tracking and pointing technology developed for surveillance 'may prove valuable in commercial aircraft guidance and in ground traffic monitoring'. He also sees potential applications for very precise sensing and measurement technology in automated process control.

SDI program managers also expect numerous commercial spin-offs from laser weapons. Rankine envisages SDI research as contributing to the use of monochromatic radiation in CT-scanners, which would reduce the dosage of radiation necessary for diagnosis while providing higher resolution. This 'would allow discrimination between molecular species, not just between density variations'. He also sees potential for laser etching of microelectronic components. Other likely spin-offs suggested by SDI staff, researchers in national laboratories and defense contractors would benefit such areas as energy generation and storage in space, super-computer hardware, software for artificial intelligence systems, and hybrid optical-electronic computers. At the same time, SDI officials warn against expecting too many specific products to emerge from the research. 'People miss the point when they emphasize widgets', says SDI chief scientist Yonas. 'The real commodity is the knowledge base, whose benefits are indirect but still real'. Colonel Joseph Rougeau, director of SDI's Office of Educational and Civil Applications, stresses that the program is taking a 'technology-push' approach to commercial applications; 'We will develop the technology we need and publicize what seems commercially useful', he says. 'The applications are limited only by the imagination and creativity of American companies' (Finneran, 1986).

Perhaps the SDI Office of Education and Civil Applications will learn from the experience of the British Defence Technology Enterprises (DTE). The DTE is one of the initiatives the British Government has taken to encourage the civil application of the results of military R&D. The government set up joint MoD-industry committees to inform industry about the program of defense research. But the House of Commons Select Committee on Defence later found that these committees had chiefly involved only the major contractors. The then Department of Industry also felt that the MoD was not doing enough to use its massive public purchasing power to stimulate industrial innovation. Then, in 1983, the Maddock report to the Electronics Economic Development Council argued that too little was being done within the major electronics defense contractors to encourage spin-off. The House of Lords Select Committee on Science spoke of the creation of a circle of large firms which had prospered through MoD contracts and which kept largely to themselves the R&D associated with this work.

By the end of 1983, the government, prompted by the Prime Minister, had begun to take steps to deal with the problem of firms 'sitting on' the results of military R&D. It had also launched a scheme which led to the establishment in December 1984 of an organisation called Defence Technology Enterprises (DTE). DTE is funded by a consortium of venture capitalists who will gain special access to the Royal Signals and Radar Establishment at Malvern, the Royal Aircraft Establishment at Farnborough and the Admiralty Research Establishment at Portsdown. The idea is that they will be able to pick out commercially applicable ideas which can be cleared on grounds of security and commercial sensitivity. They then communicate these ideas to their (much larger number of) associate members, possibly arranging further venture capital to help the projects on the way. The MoD has a share in the profits from DTE and receives a licence fee for all ideas transferred through DTE (Gummet, 1986).

In a *Financial Times* article on the same subject the author states that more ideas from the United Kingdom defense research community would reach the commercial marketplace if individual researchers moved job more frequently. According to this article, the large job mobility in the United States leads more defense scientists to seek work in commercial companies, resulting in the better United States record in transferring military innovation to the civilian sector (Marsh, 1986).

The pro-spin-off arguments are for the most part based on simply stated expectations. Empirical evidence to support these arguments however, hardly exists. Patent statistics which give information on the output of military R&D spending and its commercial potential, do not exist. Civilian spin-off from military spending has become increasingly difficult to detect as defense has moved into space. The requirements for operating in space are very different from those on earth, for example, extreme lightness, miniaturization and reliability are of much less value on earth than in space (Thurow, 1986).

The construction of military systems is a qualitative step in the development of the technologies involved: these systems, which are military end products, acquire technological characteristics which make them unfit for non-military use and certainly not in consumer goods. The growing complexity of military systems endorses this difference even more. Military research produces goods with high 'users-value': they must be designed and built to function reliably in the extreme conditions of the battle, which include extremes of temperature, vibration, shock, incident radiation, humidity and pressure. There are no really comparable requirements placed on products intended for the civilian market except in a few limited cases. No-one realistically expects a commercial typewriter that has been frozen, thrown against a concrete wall, and soaked in water to function properly, yet military hardware must be capable of withstanding such abuse. To achieve such reliability is extremely expensive and would create prices far beyond the reach of civilian markets (Tirman, 1984).

The development of very big lasers is interesting only for a limited number of exotic applications. These installations have a length of a few hundred meters and they use, because of their low efficiency, an amount of energy which corresponds with the total energy use of a middle-size city. Very big laser systems have, up till now, only been used in military projects and developments such as the 'nuclear-bomb-pumped-röntgenlaser' have definitively no civil application.

Another example of a project having limited civilian potential is the super-fast chip research program (VHSIC). The Defense Advanced Research Project Agency (DARPA) of the DoD requires that all chips for operational applications must be resistent to a radiation dose of 10^6 rad, which is the radiation intensity of a nuclear weapon of ten kilotons exploded at a distance of fifty meters. These extreme circumstances are not usual in the civilian market!

Another thing is that the civil micro chip-producers are not even interested in participating in the DoD VHSIC-program. After the first round for candidate-SDI-contractors all civil companies withdrew. Intel, the biggest supplier of chips for military use, preferred not to compete for military contracts. In the information technology sector particularly, the civil market outruns the defense market. The budget for civil R&D is so large, that research spending in military circuit looks pale in comparison. This makes the civil chipmakers insensible for seduction by the military. In the microelectronic industry the civil sector operates autonomously and is far ahead of the military sector. One might say here that the spin-off direction is reversed.

A further point is that military technology is too complex for the consumer goods market. The construction of extremely sensitive detection-systems in space is so complex that more 'down-to-earth' applications do not exist. The most optimistic estimate is that at most only five per cent of industrial robots will be able to use these advanced detection systems in 1995. The Pentagon is interested in the technical specifications of the end product. It is, of course, not relevant to the civilian sector if this means that these products must be assembled by hand. Japanese firms concentrate on the perfection of their production technology to achieve cheaper and more reliable mass production. In 1983 the American Machine Builders Association concluded that military projects had been too much concentrated on technical specifications, which led to the development of expensive and overly complex machinery.

Although the American army had already ordered the development of numerically controlled machines in the early 1960s, the market is now dominated by the Japanese who started a civilian program ten years later. In 1982 the American Semiconductor Industry Association (ASIA) placed considerable emphasis on highlighting the United States deteriorating trading position compared with Japan. According to the ASIA, this was mainly due to the fact that U.S. military programes led to indifference with respect to civil market needs. It is interesting to compare this situation with that in Japan. In the heyday of Japanese militarism just before the Second World War, the military share in production averaged between 30 and 50 per cent. In 1980 this had decreased to 0.36 per cent. The Japanese Government currently spends 2.4 per cent of its R &D budget on military goals. In the United States the equivalent figure is 48 per cent. In the words of Lee Iacocca, President of the Chrysler corporation, 'How can we compete with a country that spends only eighty dollars a year per head on military R&D while we spend more than tenfold?'

We can conclude that through the relative insensitivity of military technology-sector to commercial market needs, the United States has not been in a position to produce cheap, simple and reliable products although it has very high levels of military R &D expenditure. The extremely stringent requirements placed on military goods makes military R&D and production extremely expensive. According to one estimate by SIPRI, R &D for a military product costs on average twenty times as much as that for a civil product (SIPRI, 1981). Even if we assume that some compensation for these high development costs is included in government contracts for R &D, the high production costs of a product that meets the needs of military demands, while also having commercial applications, would still limit its consumer appeal. This is the reason why firms in the defense industry have little interest in transferring the results from military R &D into civilian products.

In estimates of the commercial application of military R&D, the problem of whether market need is sufficient to repay production costs is generally not considered: only the application possibilities are mentioned. This means that people usually overestimate the civil potential of military R &D. Demand site requirements, at least in the case of civil demand, are simply not part of the equation. Technical spin-off potential is only realized when it is applied to meet a market demand with sufficient potential to yield profits. (Eisbach, Rilling, 1985)

Measuring spin-off

Except for the above more technical arguments which illustrate that civil applications are not always as obvious as they might seem, there is other supporting material to question the spin-off argument. Perhaps the most direct and convincing means of measuring spin-off potential is to measure the spin-offs from past military projects. People often refer to the wartime experience. There is no doubt, for example, that a number of extremely valuable civil technologies, ranging from radar to penicillin, were brought to maturity in a far shorter period under the intense stimulus of the Second World War than would have been the case under peace-time conditions. But these conditions were exceptional. It is not obvious that the competitive pressure for innovation today is actually greater in military technology than it is in many consumer products in international trade. Nevertheless, there are many examples of military technical developments which have diffused into civil applications. Among the better known of these are nuclear power, liquid crystals, aero-engine propulsion, carbon fibres and thermal imaging. The question then is whether it would not have been more efficient to develop these techniques and materials directly and not via the military route.

Let us attempt to obtain a better understanding of what spin-off is and to see how it can be measured. In general, spin-off can be distinguished in three different aspects:

- the transfer and exploitation of products, technical processes or materials in areas of interest that are separated from the areas for which the technology originally was developed;
- the transfer of technical and scientific knowledge to other areas;
- the transfer of the organization and management techniques of big military or government research projects. (Schumann, 1967).

We are most interested in the second aspect, the transfer of scientific and technological information, which is the most material form of spin-off. The other two are becoming less and less relevant.

One, partial method to catch this form of spin-off quantitatively is the analysis of patent development. Using this analysis, it is possible to begin at the level of cited technological possibilities and then to count the commercially realized spin-off projects. In the United States in the early 1960s, when defense and space research played an important role in the national science system, there was a series of such quantitative studies which reached the following conclusions:

— Of the 328 patents which resulted from the NASA research programs up to 1963, only 16 per cent, according to NASA publications, had commercial potential.
— Interviews with a hundred or so government R&D contractors in the military sector, and twenty smaller enterprises showed that between 1949 and 1959, these companies filed in total 61,300 patents, 7,988 of which resulted from government-funded military R&D projects. Significantly only 7 per cent of these patents were commercialized.
— According to a second study, only 13 per cent of patents resulting from government-funded R&D projects were commercialized. (On the contrary: 50 per cent of the total of American patents is used.)

Commenting on the period from 1949 to 1959, 15 of the biggest United States companies who performed government-funded R&D announced, 'with an input of 7.6 billion dollars for R&D with military objectives, only 2,190 patents were generated'. These companies spent in the same period 5.8 billion dollars from their own resources on industrial R&D, which resulted in the generation of 23,880 patents. The same pattern occurred in the semiconductor field during a period in which military research played a major role. Only five of the 112 patents registered by Texas Instruments resulted from government funded research, and only two of these were subsequently commercialized (Rilling, 1985).

The results of the variety of studies on the civil commercial application of military R&D can be summarized in one sentence from a UN report: 'What strikes most is not how much, but rather how few results for the civil sector have come from all military R&D input' (UN, 1978). This opinion is shared by the Department of Research and Technology of the Federal Republic of Germany

(BMFT). For assessing the commercial spin-off from military programs, they use a basic 'rule of thumb' of one to ten (Bertram, 1985).

What is most striking from the list of the top ten contractors of the SDI program is that the civil sector is hardly participating at all in SDI research and only Boeing has a significant involvement. According to van Bentum, physicist and Dutch student of SDI, this lack of civilian participation is not accidental; it has to do with the growing demarcation between the civil and military sectors in the United States. The United States defense industry must more and more be regarded as a direct extension of the Pentagon. The development and production of new weapon systems is mostly carried out inside a military industrial complex which has few connections with the civil industry sector (Van Bentum, 1986).

Economic benefits of military R&D-spending*

The assumption that federally-funded scientific research leads to economic benefits for the country has been fundamental to government science policy since the end of the Second World War. Analysts have abandoned the linear model that sees a simple progression from basic research to applied research to product development, but they still believe that scientific research plays a vital role in technological progress and consequently in economic growth. Chapter 6 described present day insights into innovation processes. Economic returns, however, are neither the sole nor the primary purpose for Governmental research spending. The advancement of knowledge and specific mission agency goals such as national security, public health, and the exploration of space are all essential parts of the rationale for national research spending.

Several trends have combined in recent years to make some policy-makers more interested in economic and other quantifiable measures of research success and benefits. Technology is becoming an essential component of economic competitiveness; National budget constraints are forcing lawmakers to re-evaluate spending and to look for ways to compare the value of widely divergent government programs; quantification of program success offers the hope for an objective measure that could simplify politically con-

*Based on the report: 'The SDI: costs, contractors and consequences'; Council on Economic Priorities, 1985.

tentious decisions about increasingly esoteric and complex scientific research. One approach to simplifying research evaluation is to view national research spending as an investment that should produce a measurable economic return.

Government R&D is justified primarily by the assumption that private industry is not likely to invest in activities such as expanding basic knowledge, which do not have direct commercial payoffs, or which are too expensive for one company to undertake alone. It is assumed that government, acting as a funder of basic research and often as a first user of new innovations, contributes to economic health by expanding the country's technological base. Numerous studies, employing a variety of methodologies, have attempted to substantiate this assumption empirically.

The space program has probably been the most carefully analysed United States government R&D effort. The National Aeronautics and Space Administration (NASA) has commissioned a variety of studies examining the economic benefits of space science. The results of these efforts, and the responses they stimulated, clearly bear on an understanding of how SDI will affect the national economy.

Chase Econometrics performed one of NASA's most important studies in 1965 (Evans, 1976). It examined the relationship between NASA's R&D expenditures and productivity growth. Through statistical testing, Chase found that NASA's R&D was strongly associated with rising productivity between 1960 and 1974. Results suggested that the discounted rate of return on NASA's R&D approached 43 per cent (Hack, 1981). In other words, for every $1 million spent by NASA in 1976, Chase predicted productivity improvements resulting from the expenditure would enhance economic output by $23 million in 1984. This is clearly quite a high rate of return on R&D investment. Indeed, a U.S. General Accounting Office (GAO) review of the study noted that: 'This conclusion, if accepted, would indicate an enormous impact on the economy. The literal interpretation is that, if NASA had spent nothing on R&D, productivity would have declined in the United States between 1965 and 1974. That is, NASA's R&D is given credit for all of the nation's productivity growth, and more besides.' (GAO, 1977).

The GAO examination of the Chase study did not end there, however. Government auditors repeated Chase's statistical analysis and then varied the assumptions. GAO's exploration found that Chase's findings were very sensitive to the assumptions made. For example, if the time period and variables used in the statistical testing were altered slightly, the relationship between NASA's R&D and productivity became weaker and results were statistically insignificant. GAO was careful to report that its findings were 'not "proof" that NASA's R&D had no effect on productivity'. Indeed, GAO suggested that case studies of NASA technological contributions, such as satellite communications, would better demonstrate the benefits of the space program than macroeconomic studies such as Chase's.

Although limited in their scope and implications, case studies clearly suggest NASA's R&D has generated benefits. A study by the Midwest Research Institute in 1971 found a variety of innovations derived from NASA efforts. In-house research by NASA provided the basic knowledge necessary to make geostationary satellites possible. NASA, as the first user of new innovations, also helped companies gain valuable experience developing products. Finally, NASA research assisted industry by evaluating alternative technologies, demonstrating new devices, and reducing uncertainty. Nowhere is the value of NASA efforts more evident than in the development of communications satellites. As an analyst for the Congressional Research Service explains:

> . . . by 1971 a commercial communication satellite system had been in operation for six years, while only slightly more than ten years prior to 1971 many experts had serious doubt that satellites could be placed in synchronous orbit before 1970, that satellites could survive in space and operate long enough to be economically viable, that the quality of satellite communications transmission would be acceptable, and that the cost of satellite systems would be competitive with traditional earth-based communications (Hack, 1981).

Nonetheless, while case studies provide important anecdotal evidence, the overall economic efficiency of NASA's R&D, and government research in general, remains uncertain. Econometric studies of the relationship between productivity growth and R&D at the industry level have found that while industry-financed R&D is significantly related to productivity, government-supported R&D is not statistically connected to increased economic efficiency

(Thomas, 1980; Leonard, 1971; Griliches and Lichtenberg, 1984). Moreover, industries that are major recipients of national R&D funds have the lowest productivity return on R&D from both sources.

There are at least three possible explanations for the above findings. First, Government R&D may lower the marginal cost to the firm of performing R&D and thus expand the amount of R&D that the firm can perform (Carmichael, 1981). This explanation fits with the fact that most of the government's R&D dollars are spent in R&D-intensive industries (CBO, 1984). Yet increasing just one factor of production, research, will eventually lead to lower marginal returns on that factor as the most profitable opportunities are fully explored. Still, this hypothesis suggests a mechanism by which federal R&D could stimulate productivity growth that would not have been tested by earlier studies.

A second theory is that most of the government's R&D goes for developing products for its own use, such as aircraft and communications equipment, instead of on improving the production processes for commercial goods. The distinction between product and process improvements is important. Product innovations are usually only useful in a specific market. Process innovations diffuse throughout the economy (Thomas, 1980). This explanation suggests that government research can add to basic knowledge and national procurement can help manufacturers attain economies of scale. But government development work is not a particularly efficient way to stimulate productivity growth in commercial enterprises. Another conclusion from this theory would be that defense RDT&E is a rather inefficient way to increase economic efficiency. Most of the government's development work is performed for the military, and defense products are not designed in a cost-minimizing environment.

Finally, a recent study by Frank Lichtenberg of Columbia University found, among industrial firms, that increases in national R&D were associated with reductions in company-financed research (Lichtenberg, 1984).

In a recent report Maddock has explained the paucity of civilian benefits of defense R&D in terms of industrial structural factors, and in this respect has identified four categories of company in the electronics area (Maddock, 1983). Type A companies (or divisions of companies) are more or less completely locked into the defense market, and their innovatory efforts are determined entirely by defense requirements. They lack the entrepreneurial skills to move into civilian markets and experience great difficulties in assessing (non-specified) civilian requirements. Where such companies are specialized and largely autonomous divisions of very large multi-product companies — this is the case more often than not — the military and civilian work is completely separated due to differences in approach, customer relationships and time horizons between the two markets. The two systems are culturally very different, and as a consequence, there is little internal technology transfer.

In type B companies, civilian and defense work is performed side by side, and transfer between the two was readily achieved. However, Maddock identified only a few such companies, which were specialist software houses.

Type C companies are involved mainly in producing highly sophisticated, high-technology products for civilian applications. They would sometimes become involved in military applications of their technology, but suffered problems through excessively detailed defense specifications. Military contracts, however, offer them the benefit of longer time horizons than their civilian projects, and often offer them the opportunity to acquire new technical and design skills.

Type D companies, usually small, entrepreneurial and independent, operate almost exclusively in civilian markets, producing quite unique products. In such cases their products could achieve military application where the Ministry of Defence was willing to purchase civilian items off the shelf.

Not only do the majority of procurement contracts go to firms in category A, but so too do the bulk of defense research contracts. In the case of semiconductors, for example, Dickson showed that in 1978, 79 per cent of all Ministry of Defence research contracts went to five category A firms; in 1969 the proportion was 90 per cent (Dickson, 1983). It appears that this very high concentration of R&D funding has the dual effect of featherbedding the few recipient companies and locking-in the technology to military requirements. There is little technological proliferation of the kind achieved, for example, through a system of subcontracting by the principal defense contractors in the United States. A second important factor pointed to by Dickson is that the development project specification system is such that it can often lead to the development of devices which are so specific to defense requirements, that their potential for civilian usage is small.

The discussion above indicates how inconclusive is the research on productivity and government R&D, when the studies are viewed as a whole. Clearly, the statistical evidence is mixed on the proposition that national R&D stimulates productivity growth. It would, therefore, be hasty to assume a specific government project will either help or hinder private sector technological growth. The burden of proof clearly resides with those who would justify government projects on the basis of commercial impacts. Unless the supporters of a specific project can demonstrate how it will help, we should assume that the effect is at best neutral.

The SDI and commercial technology

One of the key factors in spin-offs resulting from the Strategic Defense Initiative is the mix of basic versus applied research involved in the project. If the work is aimed at expanding general knowledge about the technologies involved, then SDI-sponsored research might be widely applicable in the private sector. If the effort is designed to develop technologies for specific uses, then fewer commercial spin-offs may result. While this dichotomy is not strictly provable, it seems intuitively true to most economists.

General Abrahamson, SDI project director, testified before Congress that the immediate objective of the Initiative was 'to conduct research on those technologies which might enable the development of defensive systems capable of intercepting ballistic missiles . . .' This description leaves the balance between basic and applied research unclear. If we look at the programs themselves, however, it seems SDI leans much more towards developing specific applications than expanding basic knowledge.

In the surveillance, acquisition, and tracking program, for example, research work will culminate in four hardware demonstrations. In the first phase of this project, optical, infra-red, and radar signatures of re-entry vehicles and penetration aids will be measured. Then technologies such as laser optical imaging, cooled infra-red sensors, and advanced radar imaging will be explored as techniques to overcome difficulties involved in discriminating and tracking these ballistic devices. Finally, prototypes will be tested for use in each of the four phases of a comprehensive strategic defense system. Clearly, much of this research is highly application-oriented.

Other major program elements of the SDI contain a similar bias. The directed energy program aims 'to bring the most promising concepts for boost and post-boost phase intercept to an equivalent technical maturity in the early 1990s.' Demonstrating that the DoD can integrate the weapon subsystems is also being considered for this program. Goals of the kinetic energy program are:

'(1) expansion of the technology base to support the development of improved and advanced weapons and
(2) development and flight demonstration of kinetic energy weapons . . .'

In the three major programs, comprising more than 90 per cent of the SDI, most of the research seems directed toward developing specific applications and a number of prototypes.

Technologies being developed also do not seem to offer broad commercial opportunities. Private use of high-energy lasers, particle beams, large optics, and infra-red sensors are not immediately obvious. Still, possible spillovers cannot be ruled out. Development of very high speed integrated circuits (VHSIC), necessary for near real-time data processing required by a strategic defense system has, as stated earlier, some limited commercial potential, and have already found commercial use. Although the size of the market for these devices is uncertain (Steinberg, 1984) this example suggests at least some of the work performed within the SDI will have commercial value. Any industrial process technologies developed to produce and support SDI technologies could also have impact in a variety of industries. The above findings of the Council on Economic Priorities (CEP) are more or less summarized by Lester C. Thurow in his first 'rule of thumb' on military spending: no defense spending can be justified on economic grounds.

Looking at the economies of Japan and the German Federal Republic anyone can see that defense spending is not at all necessary to make a civilian economy work. Japan and West-Germany have had very small military budgets since the end of the Second World War, and both have had substantially higher ratios of civilian R&D to GNP than the United States for many years. Although classification systems among countries are so different as to render comparisons extremely difficult, American federal R&D expenditures would appear to include a far smaller share of activities that could be regarded as 'Advancement of Knowledge' than government R&D expenditures for Japan, West-Germany, France or the United Kingdom.

The actual allocation of American federal R&D resources, within the postwar structure, was one that was, and is, overwhelmingly dominated by national security considerations. Although it is possible to examine the economic effects of such an allocation, its basic rationale is extra-economic in nature. According to Rosenberg, these figures do suggest, however, that, unless it can be shown that military R&D has been generating truly gigantic benefits to the civilian sector (which he doubts that can be shown), the true opportunity costs of these arrangements have been very high to the American economy. The economic growth experience of the United States in the past few decades, by comparison with Japan, Germany and other advanced industrial countries, does not obviously support the presumption that large expenditures on military R&D have improved its relative economic performance. (Rosenberg, 1985)

SDI and the market for engineering talent

The final aspect of SDI we want to mention in this respect is the possibility the project will attract the best talent away from private enterprise. During the Apollo project, many young scientists and engineers were attracted to the effort by the technological challenge it offered. The same may hold true for SDI.

CEP constructed a rough mathematical estimate on the best available data to get an answer on the question of how many engineers, scientists and technicians SDI will require (CEP, 1985). The first step was to calculate the yearly outlays resulting from the SDI budget authority. Budget authority is simply Congressional permission to obligate funds and need not be spent in the year authorized. Outlays represent the amount actually spent in a given year, or what economists call 'final demand'. Next, CEP converted outlays into SDI-required production per industry, using the average share of production of each industry in the DoD RDT&E

Table 7.1 1987 SDI employment (millions of constant 1985 dollars)

Industry	RDT &E Production $	%	SDI Prod $	Employment per billion
Radio & TV Communication Equipment	6388	10.60	846	11660
Business Services	5421	9.00	718	9894
Complete Guided Missiles	3079	5.11	408	5621
Electric Measuring Instruments	3094	5.13	410	5647
Nonprofit organization	2901	4.82	384	5296
Electronic Computing Equipment	3263	5.42	432	5956
Electronic Components, NEC	2044	3.39	271	3731
Aircraft	1610	2.67	213	2938
Wholesale Trade	1684	2.79	223	3073
Measuring & Control Instruments	1473	2.44	195	2689
Real Estate	1640	2.72	217	2994
Eating & Drinking Places	1126	1.87	149	2055
Professional Services	1115	1.85	148	2035
Electric Utilities	1019	1.69	135	1860
Blast Furnaces & Steel Mills	910	1.51	121	1660
Hotels & Lodging Places	927	1.54	123	1692
Petrol Refining & Related Products	820	1.36	109	1497
Semiconductors	886	1.47	117	1616
Inorganic & Organic Chemistry	762	1.27	101	1391
Aircraft Engines & Engine Parts	621	1.03	82	1134
Twenty Industries Total	40783	67.69	5404	74438
All Other Industries	19468	32.31	2579	35534
Total	$60251	100.00	$7983	109972

Source: Defense-Economic Impacts Modeling System (Washington, D.C.: OUSD (PA &E) Economic Analysis Division, 1984).

account. Total production is roughly twice the amount of final demand because it includes the many products and services used and incorporated into the final product or service the Pentagon contracts for. In economic terms, production includes both intermediate and final demand contributions of each industry. CEP then calculated SDI employment requirement per industry by multiplying the average number of jobs per billion dollars of RDT &E production in each industry by the estimated level of product for SDI per industry (see Table 7.1). Finally, the numbers of scientists, engineers and technicians (SET) personnel employed

Table 7.2 1987 SDI SET employment

Industry	Engineers		Scientists		Technicians		Total
	No.	%	No.	%	No.	%	Employment
Radio & TV Communication Equipment	1,745	15.00	266	2.29	1,001	8.61	11,630
Misc Business Services	337	3.41	526	5.33	456	4.62	9,869
Complete Guided Missiles	1,453	25.92	152	2.71	538	9.60	5,606
Electric Measuring Instruments	433	7.68	68	1.21	430	7.64	5,632
Nonprofit Organizations	3	0.05	15	0.29	21	0.40	5,282
Electronic Computing Equipment	706	11.88	391	6.59	559	9.41	5,940
Electronic Components, NEC	267	7.17	51	1.37	287	7.71	3,722
Aircraft	320	10.92	55	1.88	159	5.41	2,930
Wholesale Trade	26	0.84	26	0.84	61	1.98	3,066
Measuring & Control Instruments	206	7.68	32	1.21	205	7.64	2,682
Real Estate	0	0.00	0	0.00	0	0.00	2,986
Eating & Drinking Places	0	0.00	0	0.00	0	0.00	2,050
Professional Services	247	12.15	81	3.97	332	16.38	2,030
Electric Utilities	90	4.83	24	1.31	105	5.64	1,855
Blast Furnaces & Steel Mills	29	1.75	10	0.63	27	1.62	1,656
Hotels & Lodging Places	0	0.00	0	0.00	0	0.00	1,688
Petrol Refining & Related Products	81	5.45	29	1.97	50	3.36	1,493
Semiconductors	116	7.17	22	1.37	124	7.71	1,612
Inorganic & Organic Chemistry	77	5.55	57	4.08	82	5.90	1,388
Aircraft Engines & Engine Parts	124	10.92	21	1.88	58	5.14	1,131
Twenty Industry Total	6,257	8.43	1,828	2.46	4,495	6.05	74,250
All Other Industries	2,987	8.43	873	2.46	2,146	6.05	35,444
Total	9,243	8.43	2,701	2.46	6,641	6.05	109,694

Source: CEP estimates and U.S. Department of Labor, Bureau of Labor Statistics, 'Employment by Industry and Occupation, 1982 and Projected 1995 Alternatives', unpublished data.

by SDI are estimated by the average percentage of skilled employees (by category) in each industry times the total level of SDI employment in that industry (see Table 7.2).

The assumption that SDI will require the same levels of production as the average in military RDT&E might be questioned. Many of the technologies being explored for the Initiative are very similar to those developed for other major Pentagon projects. For example, battle management and targeting systems are to be much the same as in other weapons, just quicker and more accurate. The major difference between the SDI and other RDT&E projects is its greater emphasis on technological exploration rather than full scale development. This difference would tend to bias the CEP estimates of research personnel downward. Thus the number of scientists that will be involved might be underestimated and the use of engineers and technicians overestimated.

While inexact, these estimates, nonetheless, provide a basis for assessing the role the SDI will play in the labor markets for engineers, scientists, and technicians. In 1984, the SDI will use about 4,800 engineers, scientists and technicians. That figure could rise to 18,500 by 1987 (see Table 7.3). This represents substantial growth. Though, if the assumptions of the NSF OPTIM/HIGH scenario hold true, the SDI will only be a small part of overall national and Defense Department demand. In 1987, the SDI could

Table 7.3 Projected demand for SDI SET personnel: 1984–7 OPTIM/HIGH scenario (In Thousands)

			Percentage of			
			Total Requirement		Defense Requirement	
	1984	1987	1984	1987	1984	1987
Engineers	2483	9243	0.20	0.65	1.47	4.08
Scientists	665	2701	0.08	0.30	2.56	7.94
Technicians	1730	6641	0.11	0.38	1.77	5.36
Total SET	4878	18585	0.14	0.46	1.66	4.84

Source: CEP The Strategic Defense Initiative: Costs, Contractors and Consequences, New York, 1985.

Table 7.4 Projected increase in demand for SET Personnel: 1984–7 OPTIM/HIGH Scenario (In Thousands)

	Total Growth	Defense Growth	% of Total	SDI Growth	% Total	% Defense
Engineers	177	57	32.31	7	3.81	11.80
Scientists	100	8	7.86	2	2.04	25.95
Technicians	181	26	14.13	5	2.71	19.15
Total SET	459	91	19.80	14	2.99	15.10

Source: The Strategic Defense Initiative: Costs, Contractors and Consequences, New York, 1985.

use about 5 per cent of the total SET personnel employed by the Pentagon and roughly 0.5 per cent of the national total.

SDI will take up a significant portion of the growth in SET personnel, however. It could require almost 4 per cent of all new engineers between 1984 and 1987 (Table 7.4). The Defense Department, as a whole, may require roughly a third, either directly or indirectly, of all new engineers during that period. SDI could use about 12 per cent of the Defense Department increase. Most of the work done under the Initiative will be concentrated in areas requiring large numbers of computer professionals and aeronautical/astronautical engineers. SDI might, therefore, exacerbate the likely labor market problems in those fields.

These demands will come at a time when the labour markets for engineers is beginning to show signs of tightening. As Ray Stata, President of Analog Devices in Norwood, Mass. explained, 1984 was one of the toughest recruiting years in a 'long, long time'. Analog, he reported was not able to fulfill its recruiting target for electrical engineers. Some specific areas could experience particular difficulties because of the Initiative. The extremely limited number of personnel in optical and magnetic physics/engineering could prevent the Defense Department from spending all the money it would like to in these fields.

There is some indication that, on the basis of salary alone, defense contractors are able to attract the best personnel. In a survey of its membership in 1983, the Institute of Electrical and Electronic Engineers found the median salary of engineers working in aerospace and defense was significantly higher than the median engineering salary ($45,095 versus 40,000) (NSF, 1984). The median aerospace and defense industry salary ranked sixth among the 48 industries surveyed. Salaries of new graduates going into aerospace, however, are not substantially different than other fields. Career planning and placement officials at top Universities also tend to believe new graduates are more interested in civilian than military technology. For example, Robert K. Weatherall, director of career services at the Massachusetts Institute of Technology reported: 'Contrary to a frequently heard opinion, firms working for the government do not have an edge in the competition for new graduates. It may have been so in the sixties, when the symbol of high technology were Sputnik, then Apollo, and our rivals in technology were the Russians. Today the technological symbol is the computer chip and our rivals are the commercially-minded Japanese. In many students' minds, commercial high technology is where the action is'.

Even though space and defense research may not be as attractive as it was during the 1960s, SDI might still be employing talented research personnel who could otherwise develop civilian products. The CEP-authors conclude that 'SDI may be a modest program in macroeconomic terms, but it will have a significant impact at the margin on the demand for new engineers and scientists. More importantly, even though it is only in the technological exploration stage, SDI compares in size to the largest procurement programs in the Pentagon budget including the Peacekeeper MX missile and B-1B bomber. If strategic defense technologies reach full scale development, the macroeconomic impacts could be substantially greater' (CEP, 1985).

References

Bentum, J. van, 'Spin-off van Star Wars', *Intermediair*, April 25, 1986.

Bertram, C., 'Lockrufe aus Übersee', *Die Zeit*, nr. 18, April 26, 1985.

Carmichael, J., 'The effects of Mission-Oriented Public R&D-spending on Private', *Journal of Finance*, June 1981.

Council on Economic Priorities, *The Strategic Defense Initiative: Costs, Contractors & Consequences*, W.O. Hartung, R.W. De Grasse Jr., R. Nimroody, S. Doggets, J. Brugman, New York, 1985.

Congressional Budget Office, *Federal Support for R&D and Innovation*, Washington D.C.: CBO, April 1984.

Dickson, K., 'The influence of MoD funding on Semiconductor Research and Development in the U.K.,' *Research Policy*, April 1983.

Eisbach, J. and Rilling, R., 'SDI en de beperkte spin-off van militaire R&D,' *Tydschrift voor Politieke Economie*, April 1986.

Evans, M.A., *The economic Impact of NASA R&D spending*, Bala Cynwijd, Pa.: Chase Econometrics, April 1976.

Finneran, K., 'SDI spin-off: Up in the air', *High Technology*, June 1986.

General Accounting Office (U.S.), *NASA Report may Overstate the Economic Benefits of Research and Development Spending*, October 18, 1977.

Griliches, Z. and Lichtenberg, F., 'R&D and Productivity Growth at the Industry Level: Is there still a relationship?' in Z. Griliches (ed.) *R&D Patents, and Productivity*, Chicago University, Chicago Press, 1984.

Gummet, P., 'What price Military Research', *New Scientist*, June 19, 1986.

Hack, D.B., *Studies of the Economic Effects of NASA R&D*, U.S. Congress, House Committee on Science and Technology, United States Civilian Space Programs, 1958–1978, Washington D.C.: U.S. GPO, January 1981.

Leonard, W., 'Research and Development in Industrial Growth', *Journal of Political Economy*, (vol. 29, nr. 2), 1971.

Lichtenberg, F., *The relationship between Federal Contract R&D and Company R&D*, American Economics Association Papers and Proceedings, May 1984.

Maddock, I., *Civil Exploitation of Defense Technology*, Report to the Electronics EDC, NEDC1, London, February 1983.

Marsch, P., 'Militair ideas civilian use', *Financial Times*, March 15, 1986.

National Science Foundation, 1982–87, *Projected Responses of the Science, Engineering and Technical Labor market to Defense and Non-defense needs*, Washington D.C.: NSF, 1984.

Office of Technology Assessment, *Research Funding as an investment: Can we measure the returns*, Technical memorandum, April 1986.

Rilling, R., 'Welchen zivilen Nutzen hat SDI?' *Blätter für deutsche international Politik* (5), 1985.

Rosenberg, N., *A Historical Overview of The Evolution of Federal Investment in R&D Since World War II*. Paper for workshop on the federal role in R&D, November 22, 1985.

Schumann, H., 'Der zivile Nutzen Wehrtechnischen Forschung und Entwicklung, *Wehrkunde*, November 1967.

SIPRI, *Rüstungsjahrbuch 1981–1982*, Reinbek, 1981.

Steinberg, B., 'The military boost to industry', *Fortune Magazine*, April 30, 1984.

Thomas, E., *Recent Research on R&D and Productivity Growth: A Changing Relationship Between Input and Impact Indicators?* Paper for the Third Conference on Science and Technology Indicators, OECD, Headquarters, September 1984.

Thurow, L., 'The Economic Case Against Star Wars', *Technology Review*, February, March 1986.

Tirman (ed.), J., *The Militarization of High Technology*, Ballingen Publishing Company, Cambridge, Mass., 1984.

U.N., *Die Wirtschaftlichem und Sozialen, Folgen des Rüstungswettlaufs*, Bericht des Generalsekretärs der Vereinten Nationen, 1978.

8 Summary and conclusions

The program

In March 1983, President Reagan called for an intensive and comprehensive effort to define a long-term research and technology development program with the ultimate goal of eliminating the threat posed by nuclear ballistic missiles. In January 1984, the Strategic Defense Initiative (SDI) was established as a research program based on one of a number of studies initiated following Mr. Reagan's announcement. In the same time frame, the Strategic Defense Initiative Organization (SDIO) was formed as a Defense Agency to manage the DoD effort. Specifically, a comprehensive SDI program was defined to explore key technologies associated with concepts for defense against ballistic missiles with the objective to provide a further United States President and Congress with the technical knowledge required to support a decision on whether to develop and later deploy advanced strategic defensive systems.

It is the purpose of SDI to conduct research on those technologies which might enable the development of defensive systems capable of intercepting ballistic missiles after they have been launched and preventing them from hitting their targets. Much of the public focus on SDI has been centered on the promise held out by new Ballistic Missiles Defense weapon technologies, especially the so-called directed energy weapons. The other four technical areas are kinetic energy weapons; surveillance, acquisition and tracking; systems analysis and battle management; and support programs.

The cost of the 1986–90 research program is estimated at approximately \$30 billion. The cost of an operational SDI system is estimated by the U.S. Department of Defense at between \$100 and \$500 billion and between \$400 and \$800 billion by the U.S. Council on Economic Priorities.

The SDI research program is being initiated in a period where the strategic role of science and technology is being strongly emphasized with respect to both national defense and economic development. Because of the extent of the SDI research and development program, a principal question is what impact will SDI have on the industrial base? In this study we have solely concentrated on this principal question and we have not touched upon the possibly even more important aspects of the impact of SDI on national defense policy and on foreign relations policy.

Whereas SDI and its three major aspects, industrial development, defense and foreign relations have mainly been the concern of the United States, this picture changed in March 1985, when U.S. Defense Secretary Caspar Weinberger, addressing a NATO Nuclear Planning Group meeting in Luxembourg, explicitly invited the NATO countries, as well as Australia, Israel and Japan to participate in the SDI research program. At the conclusion of this meeting, the NATO defense ministers issued a unanimous declaration of support for SDI research. NATO leaders, however, have expressed concern that the research be conducted within the terms of the Anti Ballistic Missile Treaty.

The result of Weinberger's invitation has been a series of continuing bilateral discussions with many of the countries; briefings to their delegations who have come to Washington; arranging visits for groups to United States' laboratories; and, in a number of cases, visits from SDIO teams abroad. The object of these dialogues has been to address the various procedural concerns and to identify areas of SDI research for possible participation, consistent with American law, with security interests in protecting sensitive technology, and with the United States' obligations under the ABM Treaty.

Despite these steps towards European participation in the SDI research programme, five major barriers have been formulated that limit the extent of further foreign participation:

(i) The Anti-Ballistic Missile (ABM) treaty of 1972

This is a principal barrier since almost half of the SDI program (estimated $14 billion) consists of work on the development and testing of anti-missile components limited by the ABM Treaty.

(ii) The limited capabilities of foreign companies

The demanding requirements of defense against ballistic missiles places additional limitations on foreign participation. About one-third of the SDI budget, over $10 billion, will be spent on research into technologies in which foreign companies have little expertise. The United States has been doing work on ABM systems for almost thirty years spending over $50 billion and has created a substantial industrial base in anti-missile technology.

(iii) The commercial potential of some SDI technology

Although SDI is unlikely to have significant commercial benefits, foreign companies are unlikely to receive contracts with spin-off potential (estimated at $4 billion). The SDI will be developing technologies that are irrelevant to commercial application, under conditions of strict military secrecy. Most of the SDI research with potential civilian applications is already under development by United States civilian companies. There is in fact a limited demand for some of these exotic technologies, either from the commercial or military sectors.

(iv) Some of the work must be peformed in the United States

About $1 billion of SDI consists of activities, such as management support, that by their nature must be performed in the United States. Non-United States companies are unlikely to bid on such contracts, and would be unlikely to receive these contracts even if bids were made.

(v) The competitive environment

Given the above barriers, there is about $1 billion of the SDI that foreign companies would be eligible to bid for. This estimate of $1 billion is based on a list of 18 areas of opportunity that are reported to be part of the Memorandum of Understanding signed between the United States and United Kingdom governments. However, these contracts will be eagerly sought by many qualified American companies as well.

The barriers as mentioned are not a result of bureaucratic procedures but of considered United States national policy. Ample evidence has been offered to conclude that foreign companies are unlikely to receive more than $300 million in contracts over the next five years.

In the United States it is believed that mutually beneficial allied participation in SDI would be facilitated by new government-to-government agreements. In December 1985 the United States concluded a Memorandum of Understanding with the United Kingdom. Similar agreements have been signed in 1986 with the Federal Republic of Germany, Japan, Italy and with Israel. Parties which have concluded agreements or are considering doing so have special relationships with the United States concerning their defense.

The possible participation of foreign countries in a large advanced military research program — such as SDI — has meant that they are confronted with the question of the effects of such participation on the development of their industrial technological base. Insight into the issues raised by participation is all the more pressing since some countries are seemingly basing their decisions to participate on the presumed advantages for their national economy. In the United States the discussion on the civil benefits of large military and aerospace R&D programs and on the economic benefits of military expenditure is of long standing. Several parties in the US have addressed these issues over the past decades including the National Aeronautical and Space Administration (NASA), the General Accounting Office (GAO), the Council on Economic Priorities (CEP), the Office of Technology Assessment (OTA), the United Nations Organization (UNO), as well as a number of well-known economists such as J.K. Galbraith, R.B. Reich, L. Thurow, N. Rosenberg and J. Tirman. In addition a large number of studies have been performed with respect to these issues by a wide range of researchers in order to generate evidence.

Military or industrial policy?

A summary of the results of the above studies will be preceded by a brief general discussion on the role of government in the development of industrial technology and by giving some major results of innovation studies at the level of the individual firm.

The benefits currently accepted to be associated with industrial innovation and technical changes are such that governments in most industrialized countries feel that they cannot do without explicit technological innovation policies. The pressures to construct such policies have grown as a result of a number of factors, including the worldwide deterioration of the economic climate during the 1970s, changes in the international economic order, balance-of-payment problems, the availability of new technologies, competitiveness of industry, and last but not least, the general acceptance of the case long argued by many economists that innovation and often related technical change play key roles in stimulating economic development.

A balanced and integrated industrial technology policy contains three principal and interrelated elements determining innovative performance. These three elements are:

— technical opportunity including the technical and scientific infrastructure;
— the structure of the industrial sector; and
— size and structure of market demand.

It is apparent that in an approach towards reindustrialization defined as 'the structural transformation of industry into higher added value, more knowledge-intensive sectors and product groups, and the creation of major new technology-based industries serving new markets,' science and technology are necessary, although not sufficient elements: in other words they are primary enabling factors. On the supply side, the technical and scientific infrastructure should obviously be aligned with the industrial sector and its structure. This means that not only is the structure of the industrial sector itself of great importance to reindustrialization, but also its coupling with the infrastructure.

Markets are obviously of the foremost importance to the industrial sector and to its structure. In this respect markets should be looked at not only from a more traditional standpoint of size and tariff barriers, but also from a more dynamic and cultural technology-led viewpoint which includes the concepts of 'leading markets' and 'public acceptance' of new technology. The latter issue in turn is strongly interrelated with an important element of the technical and scientific infrastructure namely education.

While the case can be made for the adoption of coherent national industrial technology policies, the question has to be posed whether the industrial structural transformation process can sufficiently advance in the absence of such policies. In the case of the United States, which has long operated on *laissez-faire*, market-driven lines by large corporations working on an international level, the answer would appear to be 'yes'. This contrasts with the experience in Japan where long-term and consistent national policies have contributed to the remarkable industrial transformation of the past thirty-five years.

But it is pertinent to ask a further question: has the United States government, in addition to exercising its advantageous position of political dominance, been involved effectively in informal rather than formal policies towards industrial and technological change? The answer to this question might be a tentative 'yes'. The Department of Defense has played its part both on the supply side (R&D funding) and on the demand side (procurement). It is being suggested that military R&D spending is seriously distorting the broad thrust of American technological development along paths dictated by military requirements and away from the current and future needs of world markets for civilian goods. This contrasts sharply with trends in Japan. On the other hand there is the possibility that military R&D might be establishing some of the technological trajectories of the future. In addition, there will be civilian spin-offs and an increase in the accumulation of national expertise in areas of advanced technology and manufacturing know-how. The question is whether these benefits will, in the case of SDI, outweigh the costs.

Technological innovation

Some over-simplified early 'models' of technological innovation emphasized the causal role of scientific and technological advance and were generally linear; for example, the so-called 'technology-push' model of innovation. According to this model, discoveries in basic science lead eventually to industrial technological developments which result in a flow of new products and processes to the market place.

From the early to mid 1960s onwards, largely as the result of a growing number of empirical studies and descriptions of actual innovations, the role of felt need in innovation began to be emphasized, which led increasingly to the adoption of the linear 'need-pull' model of innovation. Here, innovations arise as the result of a perceived, and often clearly articulated market need.

During the past decade, both the pure technology-push and need-pull models of innovation have increasingly been regarded as extreme, and untypical examples of a more general process of coupling between science, technology and the marketplace. In the first case, it is obvious that 'more R&D' has not necessarily resulted in 'more innovation'. In the second case, overemphasis on market needs can result in a regime of technological incrementalism and lack of radical innovation. Moreover, the relative importance of technology-push and need-pull might vary considerably during different phases of the industry cycle.

A more representative model of industrial innovation is the so-called 'interactive model'. According to this model innovation is regarded as a logically sequential, though not necessarily continuous process, that can be sub-divided into a series of functionally separate but interacting and interdependent stages. The overall pattern of the innovation process can be thought of as a complex net of communication paths, both intra-organizational and extra-organizational, linking together the various in-house functions and linking the firm to the broader scientific and technological community and to the marketplace. In other words the process of innovation represents the confluence of technological capabilities and market needs within the framework of the innovating firm.

The literature on factors affecting the success and failure of technological innovations is quite extensive and it can be summarized that there are no magical formulae for innovatory success. It is more or less possible to sum up the results of the various studies by mentioning a number of the major success factors:

— understanding and meeting user needs;
— effective coupling with the marketplace;
— effective internal coupling;
— effective coupling with external sources of scientific and technical expertise;
— gifted and committed entrepreneurs;
— efficient manufacturing procedures;
— efficient after-sales service and user education.

The above factors are concerned with what the firm actually did or did not do during the innovative effort; in other words they are all project execution variables. To these we can add a second set of factors that outline essentially the pre-conditions for corporate innovation, i.e. that define a corporate framework which is favourable to innovation.

These factors are:

- top management commitment to, and visible support for, innovation
- the presence of a long-term corporate strategy in which innovation plays a key role
- an associated long-term commitment to major projects
- corporate flexibility and responsiveness to change
- top management acceptance of risk
- the creation of an environment within the firm in which entrepreneurship can flourish and innovations occur.

A significant fact to emerge from these innovation studies is that, outside the United States (where public procurement has been important), governments have rarely been seen to play a major determining role in either success or failure. In some instances this has undoubtedly been a function of the research methodologies employed, which have often concentrated on firm-specific factors. It might also point to a more fundamental factor, which is that innovatory success or failure, in practice, is determined largely by the actions of management. The implication of this, of course, is that in the case of specific innovation projects in firms, there are inherent limitations to what government technological innovation policies can achieve. While governments can, to be sure, create a national environment conducive to innovatory endeavours, while they can supply certain inputs to the innovation process and act as an innovation-receptive market, and while they can remove certain barriers to innovation, in the final analysis success depends on the propensities and abilities of industrial managers.

Finally, all the major studies of success and failure in technological innovation emphasize the importance of intra-firm and extra-firm linkages. They support the interactive model of innovation as a process of communication. Establishing efficient internal linkages is, of course, a prime function of management, as is establishing appropriate mechanisms for linking the firm to the marketplace and to external sources of scientific and technological expertise and advice. With regard to the latter there exists an important role for public policy in facilitating linkages between the publicly funded scientific and technological infrastructure and industry.

In summary it can be stated that although analysis of innovation processes would indicate the importance of the development of emerging technologies within the framework of SDI, there is no guarantee for economical benefits to be derived from these technologies. The success of industrial technological innovation depends only for a small — but often essential — part on the availability of the results of scientific research and on technology. Moreover in the case of SDI it is highly questionable whether the exotic nature of these technologies would lead to converging trajectories with respect to military and civil applications.

Economics of military R&D

An important issue in the debate on the pros and cons of the SDI program is the lively level of spin-off. What will be the eventual civil benefits of military spending and what is the relationship between government R&D spending in the military sector and economic growth? Countries which do not have large military R&D efforts tend to view the latter as a form of government support to industry which gives significant advantages in high technology areas. Conversely, those which have such programs tend to regard them as a necessary defense burden which may contribute some spin-off to the civil sector but may also diminish the country's competitiveness by diverting resources away from commercial objectives. Key issues in the debate on the economies of military R&D are spin-offs to the civilian sector and opportunity costs.

With respect to the SDI R&D program the spin-off issue has been discussed with regard to the type of the SDI research work, the nature of the research, the secrecy to be guarded, the industrial structure and the mobility of research personnel. In addition cases have been presented of research investigating spin-off from earlier military and aerospace programs.

The type of SDI research

The question of whether SDI R&D could potentially generate civil, commercial spin-off depends largely on the budget for fundamental research. A breakdown of the 1982 United States Federal R&D

Table 8.1 1982 US Federal R&D expenditures (%)

	Defense	Nondefense
Basic	3.2	33.7
Applied	11.0	35.3
Development	85.8	31.0
Total	100.0	100.0

Source: Congressional Budget Office, 1984

budget into defense and non-defense components is shown in Table 8.1. The largest items of the US DoD R&D budget involve the development of advanced weapons systems and the construction and testing of prototypes. Conversely, the share of the DoD R&D budget that goes to basic and applied research is, by far, the smallest of any major US government agency. Of the total SDI R&D budget for the period 1986–90 75 per cent goes to industry, 20 per cent to government laboratories and 5 per cent to universities and small business firms for 'mission-oriented basic research'. Within the SDIO, the Innovative Science and Technology Office, has been created to spend the SDI's basic-science money and pursue highly innovative, high-risk concepts that could have a revolutionary impact on the Strategic Defense Initiative. Although the Office will not limit its awards to university-based scientists and engineers, it is hoped to tap the reservoir of brilliant minds in the academic community and promote university involvement in the SDI-program. The share of university research paid for by DoD has increased from 10 per cent to 16 per cent over the 1980–6 period and is expected to further increase because of SDI. It will, however, remain a minor portion of the total volume.

Our conclusion is that although SDI is a major research program, its small percentage of basic research does not make much spin-off likely.

The nature of SDI research

The five major technologies within the SDI program are directed energy weapons; kinetic energy weapons; systems analysis and battlement management; surveillance, acquisition and tracking; and support programs.

SDI will stress the development of performance parameters that have limited relevance to the civilian world. SDI computers must be able to survive the effects of nuclear explosions, and so SDI is funding research on radiation-hardened Gallium Arsenide computer chips. Directed energy systems can also have industrial and medical uses but the military needs weapons with power outputs of many millions of Watts that will operate for a few minutes, whereas civilian tools must operate for much longer time periods at powers of only a few Watts. X-ray lasers could be used in medical research or in the production of computer chips, but SDI is funding work on X-ray lasers powered by nuclear explosions, an energy source not contemplated for civil applications.

Although past experience shows a number of spin-off cases, like nuclear energy, aero engine propulsion, carbon fibres, liquid crystals, our conclusion is that the exotic nature of SDI research will heavily limit commercial applications.

The secrecy of SDI research

In a recent background paper on microelectronics R&D the Office of Technology Assessment (OTA) put forward well the dilemma with respect to secrecy: 'keeping information about new defense technologies within the United States is a critical concern for military security; free and wide exchange of ideas and results is an equally crucial ingredient of scientific research'.

Both the limitations under the ABM treaty and the absolute desire to keep results of most SDI research secret will strongly limit civil applications.

The industrial structure

As lists of SDI contract awards show, a large percentage of the SDI budget goes to a limited number of defense contractors with relatively few activities outside of the defense and aerospace business. The funding of mainly military and aerospace contractors will limit commercial applications of SDI research.

Mobility of research personnel

Mobility of research personnel between military and commercial research activities is a factor in the process of transfer of knowledge and constitutes an important factor in the spin-off debate. Job mobility can be considered a positive factor in the spin-off process of military R&D.

Investigations into previous patterns of spin-offs

One, partial method to catch spin-off quantitatively is the analysis of patent developments. Using this analysis in the United States in the early 1960s, when defense and space research played an important role in the national science system, there was a series of such quantitative studies which reached the following conclusions:

— Of the 328 patents which resulted from the NASA research programs up to 1963, only 16 per cent had commercial potential.
— Interviews with a hundred government R&D defense contractors and twenty small enterprises showed that between 1949 and 1959, these companies filed in total 61,300 patents, 7,988 of which resulted from government-funded military R&D projects. Significantly only 7 per cent of these patents were commercialized.
— According to another study, only 13 per cent of patents resulting from government-funded R&D projects were commercialized versus 50 per cent of all American patents. In the semiconductor field during a period in which military research played a major role, only five out of 112 patents registered by Texas instruments resulted from government funded research, and only two of these were subsequently commercialized.

The results of the variety of studies on the civil, commercial, application of military R&D can be summarized in one sentence from a UN report: 'What strikes most is not how much, but rather how few results for the civil sector have come from all military R&D input'.

Government R&D and productivity growth

Studies by the United States Office of Technology Assessment (OTA) and the Council on Economic Priorities (CEP) on the economics of military R&D seem impressive and convincing. The

OTA report on research funding as an investment concludes that it is not possible to show any returns on governmental R&D expenditures, except for some applied research programs in agriculture, aeronautics and energy designed to improve industrial productivity. The authors of the Council of Economic Priorities come to the same conclusion: while case studies provide important anecdotal evidence, the overall economic efficiency of NASA R&D and government research in general, remains uncertain. Econometric studies of the relationship between productivity growth and R&D at the industry level have found that while industry-financed R&D is significantly related to productivity, government-supported R&D is not statistically connected to increased economic efficiency. Moreover, industries that are major recipients of national R&D funds have the lowest productivity return on R&D from both sources.

Two specific cases should be mentioned in this respect. In 1983 the American Machine Builders Association concluded that military projects had been too much concentrated on technical specifications, which led to the development of expensive and overly complex machinery. Although the United States Army had ordered the development of numerically controlled machines in the early 1960s, the market is now dominated by the Japanese who started a civilian program ten years later. In 1982 the American Semiconductor Industry Association placed considerable emphasis on highlighting the United States deteriorating trading position compared with Japan. According to the ASIA, this was mainly due to the fact that U.S. military programs led to indifference with respect to civil market needs.

Opportunity costs

The heavy draw on resources — both financial and personnel — by military R&D programs has made major United States economists state that no defense R&D spending can be justified on economic grounds and that unless it can be shown that military R&D has been generating truly gigantic benefits to the civilian sector (which can be doubted that can be shown), the true opportunity costs of the these arrangements have been very high to the American economy.

The economic growth experience of the United States in the past few decades, by comparison with Japan, Germany and other advanced industrial countries, does not support the presumption that large expenditures on military R&D have improved its relative economic performance.

Based on the evidence presented above our final conclusion is that SDI is primarily a military program of advanced nature. As such, spin-off to the civilian sector will take place but, as compared with other military programs, the exotic nature of SDI will both reduce the spin-off potential and extend the time-frame in which it will take place.

From a standpoint of industrial technology policy *per se* SDI is not a cost effective approach and could be defined as counter productive.

Index